Loneliness in Later Life

Also by Hamilton B. Gibson

A LITTLE OF WHAT YOU FANCY DOES YOU GOOD: Your Health in Later Life

HANS EYSENCK: The Man and his Work

HYPNOSIS IN THERAPY (*with M. Heap*)

HYPNOSIS: Its Nature and Therapeutic Uses

LOVE IN LATER LIFE

ON THE TIP OF YOUR TONGUE: Your Memory in Later Life

PAIN AND ITS CONQUEST

PSYCHOLOGY, PAIN AND ANAESTHESIA (*editor*)

THE EMOTIONAL AND SEXUAL LIVES OF OLDER PEOPLE: A Manual for Professionals

As Tony Gibson

LOVE, SEX AND POWER IN LATER LIFE

Loneliness in Later Life

Hamilton B. Gibson
Honorary Senior Research Fellow
University of Hertfordshire

Foreword by

Peter Laslett
Trinity College, Cambridge

palgrave

Published by PALGRAVE
Houndmills, Basingstoke, Hampshire RG21 6XS and
175 Fifth Avenue, New York, N. Y. 10010
Companies and representatives throughout the world

PALGRAVE is the new global academic imprint of
St. Martin's Press LLC Scholarly and Reference Division and
Palgrave Publishers Ltd (formerly Macmillan Press Ltd).

Outside North America
ISBN 0–333–92017–1 hardcover
ISBN 0–333–92018–X paperback

In North America
ISBN 0–333–92017–1

This book is printed on paper suitable for recycling and
made from fully managed and sustained forest sources.

A catalogue record for this book is available from the British Library.

Library of Congress Cataloging-in-Publication Data have been applied for.

10   9   8   7   6   5   4   3   2
09 08 07 06 05 04 03 02 01

Printed and bound in Great Britain by
Antony Rowe Ltd, Chippenham, Wiltshire

Contents

List of Tables   vii

Foreword   ix

Acknowledgements   xi

Introduction   xiii

1 What is Loneliness?   1
2 The Problems of Later Life   21
3 The Measurement of Loneliness   39
4 Loneliness in Literature   61
5 The Benefits of Solitude   91
6 Overcoming Loneliness   109

Notes   131

Appendix A: Useful Addresses   141

Appendix B: Useful Books   146

Index   148

List of Tables

3.1   Do you feel lonely? (Age differences)   44
3.2   Do you feel lonely? (Gender differences)   44
3.3   Do you feel lonely? (Marital status)   45
3.4   Do you feel lonely? (Single women)   46
3.5   Do you feel lonely? (Marital status of men)   50
3.6   Do you feel lonely? (Health status)   54
3.7   Reasons for loneliness   55

Foreword

This is one more brief, trenchant, comprehensive book adding to the succession of titles which has made H.B. Gibson into a standing authority on ageing, especially in the Third Age. The analysis he has carried out and the advice he has given on memory, love and health in later life are all in current use by older people, and loneliness is an extremely important further subject in the series. Professional and academic as he is by training and experience, what he writes, and especially what he writes here, is addressed first and foremost to individuals in the Third Age rather than to his fellow experts on the ageing process, geriatricians and gerontologists, as they call themselves. This is a book for us and I speak as in the Third Age myself and as a member of the University of the Third Age, in the U3A.

Nevertheless his fellow professionals will have to attend to his message both because of the quality of his analysis of the wide range of evidence which he deploys but also because of the importance of what I take to be his major conclusions, which seem to me to contradict accepted expert opinion. Older people, he maintains, are not in general lonelier than younger people and solitariness does not vary commensurately with age. A further position he takes, not so much as a shock to received opinion but one which will surprise many readers, goes: living alone does not entail loneliness, and putting and keeping older people in touch with their kin is not necessarily a method, or the method, of protecting them against loneliness.

If it were true that living on one's own in later life entailed being lonely, often miserably lonely, and sometimes pathologically lonely, then loneliness would indeed be a huge problem for those in the Third Age and for those concerned with their welfare. This follows from the fact that such a high proportion of all older persons are solitaries in the sense of living in one-person households. But as our author convincingly demonstrates, these individuals are, with a few exceptions, decidedly not lonely. Moreover, numbers of those who are members of larger households are in that mostly unpleasant condition. The reasons why older solitaries are in general free of loneliness are simple. They live their lives in everyday, satisfactory, intercourse with others of their own age or younger. They pursue their avocations just as the rest of society does.

Gibson cites what he calls the 'common finding that contact with kin does not reduce loneliness and enhance psychological well-being among older adults while contact with peers often does'. He admits that his conclusions on this and other topics may have been influenced by his sample which consists of members of the U3A in Cambridge, where he lives, and U3As are organizations of older people interacting with each other and creating mutually fulfilling pursuits. But it is a common finding which he cites and we must believe his implication that other researchers have found the same thing with other samples. We must accept his conclusions therefore and go on to face the fact that there is a widespread and persistent misbelief about loneliness in later life, a misbelief of the same kind as the universal dogmatic disbelief that in former times our British ancestors all lived in large, extended households where, so a further false presumption goes, no one could be lonely.*

Gibson must not be misinterpreted as playing down the existence of loneliness at any period of life in the past and in the present, or under-estimating its significance as a liability to well-being. I find his discussion of that sad condition to be impressive and of potential use to sufferers. But he also insists on the benefits of being alone for longer or shorter periods to the independent, resilient personality, especially if given to meditation or philosophizing. The book is conspicuous for the wealth of detail with which these facts and conclusions are illustrated, particularly in literature of all kinds. Even more worthwhile for us readers, especially those in the Third Age, are the incidental comments on numerous aspects of living well during the later years. He has written a valuable and highly companionable book.

Peter Laslett

* For the fallacy that all older people lived with their kin in supposedly universally large families, see 'The Insufficiency of the Family Group in the Past and in the Present', Chapter 8 of Peter Laslett, *A Fresh Map of Life: the Emergence of the Third Age* (London: Macmillan, 1989; second edition 1996), along with its references.

Acknowledgements

I am extremely indebted to all the men and women of the University of the Third Age who contributed so generously in supplying details of their personal lives, and their experiences in later life relevant to the question of loneliness. The advice that they give will be valuable to many readers of this book, some of whom may regard the hazards of the future with some trepidation. On the whole they help to dispel many of the myths about later life and encourage us to look forward to our later years with confidence and pleasure.

One of the chapters concerns different kinds of loneliness as it has been presented throughout the ages in literature, and I would like to acknowledge the help and encouragement I have received from Pat Cahn, Anne Corlett, Vita Milne and Estelle Serpell whose specialist knowledge of different periods and aspects of literature has been invaluable.

Finally, my greatest debt is to Carol Graham with whom I have discussed every aspect of this book as it has been written. Not only has she corrected and read the whole of the manuscript, and contributed much to its stylistic presentation, but it is to her that I owe many of the ideas that are developed in this book.

H.B. GIBSON

Introduction

Loneliness is very common in present-day society among perfectly normal and quite well-balanced people. In one large social survey 26 per cent of those interviewed said that they had felt 'very lonely' within the past few weeks.[1] Those who are lonely may have quite as many social contacts as those who are not, but their loneliness derives from their being less satisfied with their relationships than they would like to be. They would like to have friends and acquaintances who are not like those whom they know, and to belong to a social network that is different from any that exists in their lives. In a marital partnership one partner may be satisfied and the other lonely. Samuel Pepys wrote in his diary:

> Home, a little displeased with my wife, who, poor wretch, is troubled with her lonely life, which I know not how, without great charge, to help as yet, but I will study how to do it.[2]

They lived a busy and sociable family life, going out together quite a lot and having visitors to the house; he was certainly not lonely but she lacked his sociable disposition and had nothing to fill her life that corresponded to his active absorption in work and music.

One may speak of two kinds of loneliness: as a *trait* or a *state*;[3] we may all experience the latter when we are undergoing a temporary period of being cut off from the sort of social interactions that satisfy us, and it will disappear when we move to a more satisfactory milieu. Trait loneliness, however, is more likely to refer to the individual's basic personality; some people with the trait of loneliness may be lonely all their lives in whatever circumstances they may live.

In any instance it is difficult to say whether loneliness is due to a person's basic trait or to the state they are living in. Two men may be living in apparently identical circumstances, but one says he is satisfied and the other complains that he is lonely. If they lived under different conditions, however, their positions might be reversed.

xiii

It is usually assumed that loneliness is especially characteristic of later life, but when researchers have tried to test this assumption it has generally been found not to be true. In one survey where they compared the population studied, decade by decade, it was found that loneliness *decreased* with age.[4] A student living communally in a hall of residence and participating in the usual group activities at a university may be very lonely, whereas an elderly widow living on her own and absorbed with her own pursuits and having just a few friends, may not be lonely at all. Older people tend to live more solitary lives but they are not necessarily lonely.

In Chapter 1 the nature of loneliness is described, for it is not the same thing as solitariness, and indeed the concept of loneliness is fairly complex and the word is used by different people to mean dissimilar things. Only when we have clarified the meaning of the concept can the problems that arise from it be understood and avoided, and remedies for the lonely be recommended.

There is, of course, a sub-set of elderly people who are very lonely, and their loneliness is largely due to such factors as bereavement, poverty and ill-health. The core of this book will concern this particular problem and how it may be overcome, but Chapter 2 will be especially devoted to examining the problems of later life that may give rise to loneliness and which derive from it, for it has been pointed out that 'Loneliness causes loneliness.' Although surveys have shown that only about 15 per cent of the retired population say that they are lonely,[5] and this is quite a low percentage compared with the population in general, all of us in the later years are more vulnerable, particularly with regard to ill-health and potential bereavement, so we need to know just how to avoid becoming prey to such desolation.

It is not easy to measure loneliness, to assess its prevalence in a population and to study how to treat it. Researchers have generally relied on interviewing people, drawing conclusions from observing their life-style, and asking them to complete self-report questionnaires. I have used the latter method in my own study which is described in Chapter 3, and there the findings are related to factors such as age, health, living conditions and bereavement in the lives of the people who were the subject of the survey.

In order to understand the concept of loneliness it is useful to consider how the theme has been dealt with in literature, and this is presented in Chapter 4. Here I do not mean the sort of books that are read more commonly by the *literati* but the well-known literature that has influenced the attitudes of the 'common man'. Most people have heard of Defoe's great work *Robinson Crusoe*, and many have read it in their childhood. Later, they may realize that the description of Crusoe on his island is a philosophical attempt to discuss the loneliness of Man and how he must try to relate to his Maker. Similarly, Swift's *Gulliver's Travels*, which is also read at one level by children, deals with the same theme, and ends with a highly pessimistic tale of how Gulliver, after having lived with the all-too-perfect 'Houyhnhnms' had to be reconciled to, and acknowledge the fact that, for him, all human relationships were flawed and unsatisfactory.

Many great writers whose works have influenced our attitudes – Charles Dickens, W.S. Maugham, Thomas Hardy, Joseph Conrad, James Joyce, William Golding, Thomas Wolfe – have written at least one book of a semi-autobiographical nature in which their own personal loneliness and their attempts to come to terms with it have been the main theme. These writers have, of course, been influenced by the work of more esoteric writers who are not well-known to the 'common man', such as Immanuel Kant, Søren Kierkegaard, Arthur Schopenhauer, Henri Bergson and Marcel Proust who have written a great deal about loneliness. An interesting contemporary example of this is David Lodge's best-selling novel *Therapy*,[6] which is the story of a low-brow writer for television who tries to overcome his personal loneliness by all sorts of modern fads in therapy and womanizing, and then gets seriously interested in the philosophy of Kierkegaard, and eventually comes to some kind of accommodation with his aloneness.

Some people have written of 'the bliss of solitude', and in Chapter 5 the positive aspects of being alone without feeling lonely are discussed. The Chinese sage Ching Chow wrote, 'What fools call loneliness, wise men know as solitude', and certainly those who have a store of inner resources can appreciate a measure of solitude without becoming lonely. Many creative writers have praised the benefits of solitude; in later life people tend to have a greater amount of accumulated skills and interests, the product of their long life's experience, which is why they may not be lonely in situations which would have made them lonesome in their younger years. As it is inevitable that we will be more alone in our later years as friends die off and families become dispersed, so we should look ahead and consider just what we intend to do to meet the changed circumstances of later life. To plan to occupy the increased leisure of retirement need not imply that we should 'disengage' from society, as was once advocated by some sociologists;[7] on the contrary we should remain active in whatever ways suit us in order to remain healthy and enjoy life to the full in the Third Age,[8] although the activities in which we take part will necessarily be different from those we enjoyed when younger.

The many different kinds of activities that have become available to older people in more recent years are described in Chapter 6. This is the first time in history that the structure of the population has changed radically so that now there is a much greater proportion in the Third Age – that is, people who have lived beyond the age of full-time employment and still have the prospect of another 20–30 years of healthy, active life, as has been so graphically described by Peter Laslett.[9] With this change in the population structure there has come into being a large number of agencies and associations such as Age Concern, the University of the Third Age, and the Over Fifties Association. In Appendix A there is a list of addresses of organizations that will be useful to people in later life, and a list of useful books is given in Appendix B.

It is contended that a very great deal can be done to overcome loneliness in later life, and this involves altering the stereotype of ageing as it is perceived by the public at large and by the elderly themselves. Seminal work has already been done by a number of pioneering writers whose endeavours will be discussed, and by organizations such as Age Concern. The paradox that has emerged from research – that the elderly perceive themselves as less lonely on the whole than the young – should not blind us to the fact that a very great deal needs to be done in a practical way for the sake of the minority who have every cause to experience problems that both cause loneliness and arise from their forlorn condition.

1
What Is Loneliness?

The question of 'What is loneliness?' must be faced before any attempt to discuss its nature, causes and implications in later life can be made. Indeed, it will be seen in Chapter 3 that the various answers to the question 'Are you lonely?' imply that elderly people sometimes interpret the question rather differently. The question does not call for any dictionary-type definition, for it is obvious that there are different types of loneliness, and in later life we may experience varieties of loneliness that are not common in our earlier years.

Growing old brings us new challenges. We know that it is too late to experience certain things, but we do not know what fresh problems are in store for us in the limited span of years in the future. We know that many friends and loved ones, being mortal, will be taken from us, and the example of our older acquaintances warns us that many of our own capacities that we have taken for granted for the whole of our lives, including mobility, hearing, sight, and even memory, may fail us. To age successfully requires courage and fortitude, and a stoic acceptance that loneliness may be our lot, but strange to relate, those who have investigated loneliness in general over the span of adult life have found that instead of increasing, reported loneliness tends to decrease with age.[1] For as well as new problems there may be unexpected fulfilments, and even excitements, to be experienced in the years ahead, and we may be less lonely than was our lot in earlier periods of our lives.

The term 'loneliness' refers to both to an *experience* and a *feeling or emotion*. Thus a man may say, 'I miss my wife; I'm lonely for her while she's staying with her daughter in Scotland, but she will be back at the end of the month.' Thus an episode of loneliness may be very specific in the life of someone who is not, in general, lonely. In this case the *emotion* of loneliness may be experienced for a very limited time, and interestingly enough, what the experience felt like at the time may be rather difficult to recall in retrospect after the period of distress is over. This may be compared with the fact that some women who have had a most painful experience of childbirth will be unable to remember it later on when they are happy with their babies. Thus when researchers ask people whether they have ever been lonely, some will reply 'No' quite honestly when it is not true.

This kind of limited loneliness may be contrasted with the form which is long-lasting and characterizes a situation in which the individual feels generally forlorn and deprived of supportive and congenial company. Thus someone may say, 'I was very lonely for the whole of my fifties when I was unemployed in Canada and felt like a fish out of water. I felt as though I had come to the end of my useful life.' Such a period may be well-remembered and indeed brooded upon at times, although the exact quality of the feeling may be difficult to recapture.

Some people appear to be permanently lonely and exhibit the *trait* of loneliness, never succeeding in assuaging their deeply-felt sense of isolation wherever they are and whoever they are with. This is conveyed by the poet Yeats who wrote a great deal about loneliness and spiritual desolation, as in his dramatic poem, *The Land of Heart's Desire*:

> The wind blows out of the gates of the day;
> The wind blows over the lonely of heart,
> And the lonely of heart is withered away.

Loneliness may dominate much of a person's life as Bertrand Russell revealed in his *Autobiography*. He wrote:

> Three passions, simple but overwhelmingly strong, have governed my life: the longing for love, the search for knowledge, and unbearable pity for the suffering of mankind. These pas-

sions, like great winds, have blown me hither and thither in a wayward course, over a deep ocean of anguish, reaching to the very verge of despair.

I have sought love, first, because it brings ecstasy – ecstasy so great that I would often have sacrificed all my life for a few hours of this joy. I have sought it next, because it relieves loneliness – that terrible loneliness in which one shivering consciousness looks over the rim of the world into the cold unfathomable lifeless abyss. I have sought it, finally, because in the union of love I have seen, in a mystic miniature, the prefiguring heaven that saints and poets have imagined. This is what I sought, and though it might seem too good for human life, this is what – at last – I have found.[2]

Russell's final mention of his having found, at last, the resolution of his loneliness refers to his marriage to Edith Finch at the age of 80. This was certainly his most successful marriage and put an end to a lifetime of marriages and divorces, as well as a good number of temporary affairs. The last 18 years of his life were certainly the happiest for him, and his long struggle with loneliness that had made him so restless, appeared to have ended.

Elsewhere in his autobiography Russell wrote of an experience that he had in middle life that must be called mystical. It seemed to him a revelation of the nature of Man's essential loneliness:

Suddenly the ground seemed to give way beneath me, and I found myself in quite another region. Within five minutes I went through some such reflections as the following: the loneliness of the human soul is unendurable; nothing can penetrate it except the highest intensity of the sort of love that religious teachers have preached; whatever does not spring from this motive is harmful or at best useless; it follows that war is wrong, that a public school education is abominable, that the use of force is to be deprecated and that in human relations one should penetrate to the core of loneliness in each person and speak to that.... At the end of five minutes I became a completely different person. For a time, a sort of mystic illumination possessed me. I felt that I knew the inmost thoughts of everybody that I met in the street, and though this was, no doubt, a delusion, I did in actual fact find myself in far closer touch than previously with all my friends, and many of my acquaintances.[3]

Four types of loneliness

A number of writers have described their idea of the different types of loneliness that may be recognized. Thus the philosopher Rubin Gotesky has delineated four types of what he called 'aloneness'.[4]

1. *Physical aloneness*
This affects us all, of course, and it is debatable whether we should regard it as a type of loneliness. We are separated from all other men and women in space and we shall never meet or indeed know of the individual existence of the overwhelming majority of mankind, even if they live in the same country, town or village as ourselves. Such physical aloneness normally causes no-one any distress, although some of us, of course, may yearn to have some sort of contact with people of distant lands, and indeed, their own countrymen, and feel that their own lives' experience has been rather insular.

2. *Loneliness as a state of mind*
This state of mind consists essentially of feeling that we are rejected by our fellows and excluded from their activities and interests when we wish to be more intimate with other people and to participate in their activities. This 'sense of abandonment' is experienced by most people at some stage of their lives and can be very distressing. At the extreme it may lead to the mental disturbance of 'paranoia', a state in which the sufferer not only feels isolated from everyone else, but that they are actively hostile and engaged in malicious plots. Even in normal experience the feeling of being cut off from those with whom we wish to associate can be very traumatic and lead to greatly lowered self-esteem. This sense of alienation may apply not only to people, and society in general, but for some individuals, from God, and lead to a conviction of being very sinful.[5] A paradoxical situation may arise in which individuals who have led most upright and blameless lives become convinced in their later lives that

they are unworthy and wicked, although they find it hard to specify exactly what wrongful acts they have committed. This condition used to be referred to as 'involutional melancholia' by psychiatrists, although nowadays it is more commonly regarded as a form of late-life depression.

3. *The feeling of isolation due to a personal characteristic*
This type of loneliness is characterized by Gotesky as, 'the rational recognition that men face conditions of existence in their relation to others which they do not know how to change'. Thus, for instance, a coloured person living in a white society where there is racial prejudice may feel isolated by very real barriers to his participation in the the normal activities that go with citizenship even though he has formal legal rights. Such a person knows that he cannot do anything in the immediate present to change his lot; the best he can do is to associate mainly with others who are similarly disadvantaged, as in an invisible 'ghetto', and to work for the removal in the long run of the roots of racial prejudice in that society.

A personal characteristic that we shall all possess one day (unless we die young) is that of being an 'old person'.[6] Such an attribute may bring on this feeling of isolation if we live in a society that is geared almost entirely to the needs, habits and values of young people. Society is always changing and it may be that older people are made to feel like fish out of water in a milieu that is strange to them. Swift dealt with this aspect of ageing in his *Voyage to Laputa* where Gulliver meets with a special kind of people, the Struldbruggs, who are literally immortal and hence condemned to go on living however sick, decrepit, poor and unhappy they are. Among the disadvantages they experience is their being increasingly cut off because of the changing language. He writes:

The language of this country being always upon the flux, the Struldbruggs of one age do not understand those of another, neither are they able after two hundred years to hold any conversation (farther than by a few general words) with their neighbours the mortals, and thus they lie under the disadvantage of living like foreigners in their own country.[7]

Although this is a grossly exaggerated fantasy, there is a kernel of truth in it. Older people may indeed feel at a loss completely to understand the contemporary use of language, and words like 'gay' which have taken on a new meaning, may seem bizarre to them in everyday conversation.

According to some writers this sort of loneliness is necessary for those who engage in certain life-styles that go with creative activity in the arts or sciences. The creative person has to stand back from society in order to achieve a detached and critical viewpoint so as to contribute original ideas and work.

4. Solitude

According to Gotesky, solitude is a variety of loneliness, and in this case there is a positive value. In his words, 'solitude is that state or condition of living or working alone, in any of its many forms, without the pain of loneliness or isolation being an intrinsic component of that state or condition'.[8] While most people would agree that solitude, as here defined, has its advantages, as in the serenity achieved by sages, it may be argued that it is artificial to regard it as a type of loneliness. It will be contended in this book that the happy achievement of solitude, when it is necessary, is a way of overcoming loneliness. We are not all sages or specially creative people, but we all need to come to terms with ourselves, and while benefiting from human companionship both in giving and receiving friendship and love, we need to achieve serenity in our own company, and to develop resources so that we respect ourselves and are not over-dependent on the company of others.

Other views of loneliness

One way of regarding loneliness is to think of it entirely in terms of the degree of integration the individual has with social networks. This poses some interesting questions in this modern age with its vastly increased number of channels of communication.

It may be difficult to enter into the world of someone who is deeply involved in the world of sport. Generally this applies to men rather than women, although there are counterparts to be observed in the female world. Such a man may lead quite an

isolated life in terms of his living arrangements, the family and
neighbourly contacts that he has, and appear to be involved in
no particular social network, yet he is not lonely at all. His
interests, his emotional life are dominated by the active partici-
pants in the sports which he 'follows', and the teams they belong
to. He will be very knowledgeable about the individual athletes,
their history and their present prospects; he will know all about
the record of the various teams and how they stand in relation
to one another. He will be elated or depressed by the latest suc-
cesses and failures of the team he favours and the athletes he
admires, just as though they were members of his own family. I
have known older men who were far more involved in the ca-
reers of athletes they may be said to 'love' than in their own
grown-up children with whom they had little real contact for a
long time.

Such sport-obsessed men are by no means uncommon in later
life and they may suffer agonies of suspense when the outcome
of matches is in the balance. The enormous amount of time
devoted to sport on TV attests to this important aspect of modern life, and now it is a worldwide phenomenon. The BBC World
Service broadcasts all night, and the programme is repeatedly
interrupted by the Sports Round-Up. Through this channel there
must be thousands or perhaps millions of people involved with
one another, albeit in a passive manner, in a world-wide network of sports enthusiasts. The BBC ran a series of programmes
in which the actor Michael Palin went on a trip almost from
pole to pole, interviewing people in many distant lands. When
he was in a remote village in Africa he met a villager who spoke
passable English and he revealed in their conversation that they
were both enthusiasts for the soccer team Manchester United.
Radio and TV have extended our social networks very far!

While men are undoubtedly the main enthusiasts for sport,
there is some evidence that women are more involved in the
huge 'extended families' of long-running soap operas. It was after
the war that the BBC started the programme *The Archers* and it
has been so popular that it has been continued ever since. It
soon became apparent that an enormous number of people were
taking it very seriously indeed, some people half-convincing
themselves that the radio characters were real, and that real-life

family dramas were in progress. Presents were sent through the post to the fictional characters, and when one of the characters (Grace Archer) had to be killed off because the actress could no longer continue in the series, there was real distress experienced by many individuals who reacted as though a member of their family had died.

The phenomenon of lonely people living a fantasy life involved with the fortunes of fictional characters is not entirely new. In the nineteenth century when Charles Dickens was publishing his immensely popular *Household Words*, a monthly magazine, for a long period each issue contained a chapter of whatever novel he was engaged in writing. While he was writing *Dombey and Son* many of his devoted readers realized that Paul Dombey, the sickly little boy who had captured their hearts, was failing in health, and Dickens received many letters imploring him not to let Paul die. When Dickens, in his own words, 'slaughtered' him, according to one commentator, 'The nation was plunged into mourning.'[9]

In the modern age radio and TV have a much wider influence than any printed material could have since they appeal to the least literate sections of the populace. These channels of the media provide an interesting example of how our highly technological society, having created a great deal of loneliness and *anomie* by breaking down the old social order, has created something that is a partial substitute for actual social relationships and provided a palliative against widespread loneliness.

Fantasy loneliness

Ronald Rolheiser, a well-known cleric and writer in Canada, has proposed several types of loneliness which he discusses from a religious point of view. Many of them overlap those which have already been described and discussed, but the variety he calls 'Fantasy loneliness' may have some special application to older people. He writes:

> Now all of us, according to degree, live in fantasy and delusion, not quite in tune with reality. We live with certain fantasies and illusions of who we are and where we fit into reality. We daydream, and after a while get part of our dreams and fanta-

sies mixed into how we see and interpret reality. Sometimes our assessment of ourselves and our place in life is close to reality; at other times it is fraught with illusion and unreality. To the degree that we are not truly and totally in touch with reality as it is, we are alienated and lonely.[10]

Fantasy loneliness may be engendered in people after retirement when they have depended very much for their self-image on the status they have enjoyed for many years, say, as head of an important department. After retirement they find that no one treats them with the deference to which they have become accustomed, and no one is particularly impressed when they give their opinions and unwanted advice. The sensible course of action for such people when they still have a great deal of energy in retirement is for them to get involved in voluntary work and apply their management skills to useful ends, and perhaps again become a big fish in a rather smaller pond. But some people are unable or unwilling to make this change, and they resort to creating an unreal world in which they are still an important figure, and spend a great deal of time and energy in trying to convince others that they still hold positions of power and influence. To some degree they may succeed in convincing themselves.

The loneliness of such imposters is pitiable, for inevitably they become tedious bores whose company is shunned by other people, and being shunned they become even more lonely and then redouble their efforts to build up the fantasy. One such case was publicized some years ago, possibly with some journalistic embellishments. It concerned a man who had been an important figure in the world of commerce, and after retirement went to live in a pleasant village. Being separated from his wife and having no immediate family or circle of friends in his locality, he took to going to the pub to increase his social contacts. The village had two pubs, and when he became a tedious bore in the one he frequented, customers tended to desert it and foregather in the other one. When he found himself with very few people to talk to in the bar he switched his patronage, with the result that the second pub emptied and the first one gained the custom. He became the bane of the two landlords, acting as a Jonah whose presence they feared, for the locals would peer in the bar to see if the coast was clear before they entered.

This poor man's fantasy was that he was an important authority on local history, and he was reduced to picking on strangers and holding them as a captive audience while he tried to impress them with his local knowledge and to give them the impression that he was really the squire of the district, a detail that was very far from the truth.

We have all probably known men and women whose loneliness was self-evident and who have embarrassed us by trying to make us a party to their fantasies. We may feel guilty at shunning their presence and thus contributing to their loneliness, and in a shamefaced way may go along with them, to some extent pretending to believe that they really do know the important personages whose names they drop, and that they are members of this and that prestigious committee.

The loneliness of the handicapped

A minority of people are prone to a degree of loneliness for the whole of their lives because they are born with some physical or mental handicap which makes it difficult for them to relate to other people in a normal way.[11] The extent to which handicapped people manage to overcome their disability depends to a great degree on how they are regarded and treated by society. The treatment of the unfortunate used to be very much harsher than it is today; in village life there was a greater or lesser degree of tolerance and care of those who were abnormal, although it used to be accepted that the 'village idiot' was the natural butt of crude and cruel humour. With growing urbanization the handicapped were likely to be shut away in large 'mental deficiency' institutions, little distinction often being made between those who were of very low intelligence and those who were emotionally disturbed with conditions such as schizophrenic disorder. One reason for sequestering the handicapped from society was to prevent them from breeding and thus running the increased risk of producing handicapped children who in their turn would be a charge upon the public purse. This practice was widely abused, and medical officers occasionally certified uneducated girls of normal intelligence as 'mentally defective' simply because they had had more than one illegitimate child, and the certification was to prevent them having any more.

The plight of the congenitally handicapped should make us aware that the loneliness it engenders may be our fate in later life, and the loneliness may be more acute because the elderly person has had no experience in dealing with a particular handicap. Someone who has been blind or partially sighted from birth will have acquired all sorts of techniques such as learning to read Braille and getting around without the aid of sight, but when sight begins to fail seriously in the later decades it may come as a catastrophe and utterly disrupt the sufferer's way of life. It is the same with deafness: those who are born deaf learn to communicate by sign language and can form excellent social networks with other deaf people. When deafness strikes in later life the elderly deaf person may become more and more cut off from social converse, and lonely. In the survey described in Chapter 3 several people attributed their loneliness to their deafness. Even with a hearing-aid it is often impossible to hear what is being said at any social gathering because the ear may lose its power to discriminate between different channels of sound, so all that can be heard at a party is a meaningless blur of noise. For some reason deafness has always attracted some stigma and old people with their ear-trumpets were a stock figure of fun. Now that we have electronic hearing aids the position is better, but some people are still ashamed to wear them, preferring to remain in a lonely prison of partial silence than to be seen wearing an instrument in the ear.

According to Alex Comfort:

In Britain where hearing-aids are available to seniors without charge on prescription, only 400,000 out of 1,300,000 deaf people over sixty-five have them. There is still much unrelieved and even unassessed deafness. An estimate for 1962 suggested that a quarter of a million older people had severe deafness, but no hearing-aid, while perhaps one and a half million deaf people had not even had an aural examination within five years. Either they regard deafness as a natural part of ageing or they aren't aware that treatment is possible, and the same appears to be true of some doctors.[12]

It should be noted that although the hearing-aids obtainable on the National Health are better than nothing, they leave a lot to be desired. A good instrument which may be obtained from a private hearing consultant may cost nearly £1000, and this is one of the many instances which underlines the fact that in old age there is a growing gap between the quality of life of the affluent and the poor. But even a sophisticated and expensive hearing aid will not restore a deaf person's hearing to normal – they remain to some degree in a lonely prison of incomplete communication. It is embarrassing to have to repeat, 'What's that you said? Please speak up – I can't hear you.' Deafness often appears to the outsider as slowness of wit, and it is no wonder that many older people are regarded as a bit stupid, and left out of social occasions.

Another handicap that is all too common in later life and leads to loneliness is restriction of mobility which at the extreme can render people housebound.[13] One of the conditions that restricts mobility is arthritis, which is a degenerative disease of the joints, and as this is relieved by rest it is natural that sufferers, who used to get out and about a great deal, have to restrict their activities more and more. The use of a car is a great boon in such conditions, and in severe conditions, a motorized wheel-chair; here is another instance of poorer people being specially penalized in later life. Only when we are deprived of mobility do we realize how much in modern life, with its small families and scattered social networks, does our contact with other people depend upon mobility.

Public transport systems do, of course, give concessions to pensioners, but this means that the vehicles have to be designed so that handicapped people can get on and off them without hazard, and any elderly person who has travelled on the London Underground Railway system carrying or wheeling a suitcase is made to realize how many flights of stairs there are to negotiate. There is no avoidance of the fact that even with modern prostheses such as artificial hips and knees, reduced mobility may reduce elderly people's contact with others to the extent that many have to become resigned to a degree of loneliness.

Loneliness as a necessary driving force

There are those who regard loneliness as inevitable in civilized society, and indeed, the basic force that produces culture. Many animals naturally live in a group, and indeed could not survive for very long in the harsh conditions of nature if they were separated from their herd, pack or flock. Separate a wild animal and it becomes instinctively lonely and seeks its fellows. There are exceptions, of course, as among the large carnivores who lead a solitary life for most of their time when they are adult, although they associate in family groups. Human beings are essentially weak as individuals compared with other animals, but they have attained their dominance over all other species by forming close cooperative bonds, and through the development of language human society and culture has emerged.

The paradox is that creative and innovative individuals are often people who are by nature and circumstances somewhat lonely, and reacting to their loneliness by making sustained efforts, whether they are artists or scientists, to achieve new forms which are socially meaningful. No musician is content to create music which is never heard; no writer wants to write books that are not read; no scientist is content to produce theories and artefacts that are not socially meaningful and accepted. By their work creative people overcome the pain of their loneliness by working on projects that integrate themselves with society, producing that which is socially recognized as valuable and interesting. To fail as an artist or scientist is to remain unrecognized and therefore to remain lonely. The lonely lives that many highly creative people have led is a price they were willing to pay for achieving their ultimate goal of recognition, even when they feared that their work would be appreciated and accepted by future generations only after their death. The American jurist Oliver Wendell Holmes expressed it thus:

> Only when you have worked alone – when you have felt around you a black gulf of solitude more isolating than that which surrounds the dying man, and in hope and in despair have trusted to your own unshaken will – then only will you have achieved. Thus only can you gain the secret isolated joy of

the thinker, who knows that a hundred years after he is dead and forgotten men who have never heard of him will be moving to the measure of his thought.[14]

Loneliness and sexuality

Since time immemorial people have wondered about one of the great mysteries of creation that applies to both animal and plant life: whatever the creature, it appears in two forms, the male and the female. It seems that we, and every other living being, were born incomplete, and the life-force that continues the existence of life on earth depends upon an essential loneliness being resolved – the urge to mate. Different cultures have given rise to legends that sought to explain this great mystery in various ways.

The Jewish culture, being very male-dominated, explained the mystery thus in the Book of Genesis:

> And the Lord God said, It is not good that man should be alone; I will make him an help meet for him. . . .
> And the Lord God caused a deep sleep to fall upon Adam, and he slept; and he took one of his ribs, and closed up the flesh instead thereof;
> And the rib, which the Lord God had taken from man, made he a woman, and brought her unto the man.[15]

Thus in this legend it appears that originally there was just the one sex, the male, and it was only later that the female was created more or less as an afterthought, to be an assistant to the man and relieve him of his loneliness.

The Greek civilization gave rise to a most interesting legend which is commented on by Plato in his *Symposium*. The legend was that originally there was a human race of hermaphrodites which became split in half by Zeus's anger, and thus became humans as we know them; Plato quotes Aristophanes' speech in praise of human love:

> when the work of bisection was complete it left each half with a desparate yearning for the other, and they ran together and flung their arms around each other's necks, and asked for nothing better than to be rolled into one. . . . And whenever

one half was left alone by the death of its mate, it wandered about questing and clasping in the hope of finding a spare half-woman – or a whole woman as we should call her nowadays – or a half-man. And so the race was dying out. . . . So you see, gentlemen, how far back we can trace our innate love for one another, and how this love is always trying to reintegrate our former nature, to make two into one, and to bridge the gulf between one human being and another.[16]

Hindu culture has given rise to a legend which presupposed an original race of hermaphrodites who were over-large, and it explicitly emphasizes the principle of loneliness rather than sexuality. In the *Upanishads* we read:

Therefore a man who is lonely feels no delight. He wishes for a second. He was so large as man-and-wife together. He then made his Self to fall in two, and then arose, husband and wife. Therefore Yagnvalkya said: 'We two are thus (each of us) like half a shell'. Therefore the void which was there, is filled by the wife.[17]

The idea that one may be less lonely as an individual by splitting into two and having one's other half as a companion is intriguing.

Certainly, although the sexual urge to mate is a strong one, it is not the only reason why males and females form couples, and for that matter, homosexual men and women form couples with members of their own sex. In later life it is unfortunate that women live six or seven years longer than men, for it means that the further we go up the age-scale there will be a larger and larger preponderance of widows and other unattached women who have very little prospect of forming a new relationship with a man. How far this numerical imbalance between the sexes in the later years of life is responsible for loneliness will be examined in the research that is reported in Chapter 3. The generations who are now in the Third Age are, of course, strongly influenced by the customs and taboos of earlier ages, including the taboos against homosexual relationships which are regarded more liberally by younger people. There are already signs that more unattached older women who have previously been heterosexual for the whole of their previous lives are now forming partnerships with other women.[18]

Existential loneliness and religion

It is rather difficult to explain what is meant by existential loneliness, but it certainly accounts for a feeling of unease, dissatisfaction with life and with the self, that is felt by many people even though by all the usual criteria we would expect them to be quite fulfilled and content. A man or woman may be apparently happily married, with children, a circle of friends and an interesting job, yet feel strangely lonely. Existential loneliness may lead to conditions of depression, conditions which are very difficult to treat because there does not appear to be anything particularly wrong in the patient's life, and such mental sickness may occur at any phase of life, from adolescence to old age.[19]

The essence of existential loneliness in old age is sometimes described well by creative writers: E.M. Forster describes the condition in his novel *A Passage to India* where the old lady Mrs Moore begins to act strangely and to reject the usual family ties, becoming contemptuous of the values of her son and unpitying of the dilemma of her disturbed protégée Adela Quested. She came out from India to see her son, a magistrate under the British Raj, together with her young friend Adela, who was engaged to him. She was a benign old lady not quite knowing what to expect, but her experience of British India and an unpleasant incident at a visit to the Marabar caves had a profound effect on her and she ceased to be benign. She appeared to lose her Christian faith and to lean towards a rather sterile Eastern mysticism which made her wish to withdraw from her family and their concerns. Talking to her son and Adela she burst out:

> 'Say, say, say,' said the old lady bitterly. 'As if anything can be said! I've spent my life in saying or listening to sayings; I've listened too much. It is time I was left in peace. Not to die,' she added sourly. 'No doubt you expect me to die, but when I have seen you and Ronny married, and seen the other two and whether they want to be married – I'll retire then into a cave of my own.' She smiled, to bring down her re-

mark into ordinary life and thus add to its bitterness. 'Some-where where no young people will come asking questions and expecting answers. Some shelf.'[20]

The mention of 'a cave' is of course a reference to the grim Marabar caves where Adela is supposed to have been assaulted, and the echo makes all statements a meaningless noise – bou-uom. Mrs Moore's experience is described:

> The crush and smells she could forget but the echo began in some indescribable way to undermine her hold on life. Coming at a moment when she chanced to be fatigued it had man-aged to murmur, 'Pathos, piety, courage – they exist but are identical, and so is filth. Everything exists, nothing has value.' If one had spoken vileness in that place, or quoted lofty po-etry the comment would have been the same – 'Ou-boum'. If one had spoken with the tongues of angels and pleaded for all the unhappiness and misunderstanding in the world, past, present, and to come, for all the misery men must undergo whatever their opinion and position, and however much they dodge and bluff – it would amount to the same, the serpent would descend and return to the ceiling. Devils are of the North, and poems can be written about them, but no one could romanticize the Marabar because it robbed infinity and eternity of their vastness, the only quality that accommodates them to mankind.[21]

This fictional account strikes a true note because some individuals do towards the end of their lives experience a loss of faith in the values by which they have always lived, an existential loneliness. Whether it comes suddenly as in the case of Mrs Moore, or develops slowly over the years, the result is much the same. Loneliness is both experienced and to some extent embraced as an alternative to the social world with all its petty shams and convenient pre-tensions. We may well ask, 'And where does religion relate to all this?' The Christian view is in the words of St Augustine in his *Confessions*, 'You arouse him to take joy in praising You, for You have made us for Yourself, and our heart is restless until it rests in You.'[22] In other words, loneliness is simply a thirst for God.

The theological writer Ronald Rolheiser describes one category of loneliness as 'Restless-loneliness' which is very like the existential variety. Being a Christian he agrees that loneliness is a thirst for God. He says that 'this type of loneliness is not caused directly by our alienation from others, but from the very way our hearts are built, from our structure as human beings'.[23] I would not agree that human nature necessarily condemns us to experience existential loneliness, nor that such a condition is 'a thirst for God'. Christians are apt to maintain that in later life men and women turn to religion as the span of their future years shortens and they consider their eventual death. There is really no good evidence that this is so; there are individual cases that support this idea, but these are balanced by the cases of people who have led conventionally religious lives, and then in their later years turn away from religion, and sometimes bitterly regret the time and effort they used to put into practices that they now regard as fatuous and even demeaning. Older people may look back on their past and consider that their former religious beliefs have impoverished rather than enriched their lives; they recall the deeds committed and opportunities lost because of religious principles which now appear to them priggish and ridiculous. A sense that life has been wasted may powerfully contribute to existential loneliness in later life.

Those who have always lived by humanist principles and are atheist in belief have always expected mere annihilation at the end of their lives; they have nothing to lose if they have expected nothing. It is those who change their beliefs radically who have to face what Bertrand Russell called 'that terrible loneliness in which one shivering consciousness looks over the rim of the world into the cold unfathomable lifeless abyss'. Such a one is the well-known writer A.N. Wilson who for many years was an ardent Christian apologist. In middle-age he found that intellectually he could no longer sustain his beliefs and wrote, 'It is said in the Bible that the love of money is the root of all evil. It might be truer to say that the love of God is the root of all evil. Religion is the tragedy of mankind.' This is how he begins his book *Against Religion: Why We Should Try to Live Without It*,[24] in which he castigates not only Christianity but all religions. It is all very well for those who change their belief systems radically in middle-age; they have half a lifetime to build up an alternative system. It is those who make the change late in life who are at risk of feeling stranded, abandoned, forlorn – lonely even though surrounded by family, friends and familiar acquaintances.

The term 'existential loneliness' is used rather differently by different philosophers and other writers. The truly 'existential' philosophers, of whom Jean-Paul Sartre is perhaps the best known by the general public, teach that we should learn to use our loneliness positively; we should recognize it and not try to deny it. Interestingly enough most inquiries into loneliness have been among adolescents who are commonly searching for meaning, identity, and how they should relate to society. The popular writer Clark Moustakas holds that loneliness can be a creative force for adolescents and states:

> Every real experience of loneliness involves a confrontation or an encounter with oneself. The encounter is a joyous experience. . . . Both the encounter and the confrontation are ways of advancing life and coming alive in a relatively stagnant world. They are ways of breaking the uniform cycles of behaviour.[25]

While existential loneliness can be a creative force at the beginning of life since it can lead adolescents to face the fact that they have to think about how they can best fit into society, it is more problematic how useful it can be at the other end of life. Traditionally 'the old' should stoically resign themselves to the fact their earthly existence will soon be over and they should prepare to meet their Maker. This sad and traditional view of later life is expressed in the biblical book of *Ecclesiastes* where it is presented in poetic form:

> Remember also thy Creator in the days of thy youth,
> Or ever the evil days come,
> And the years draw nigh, when thou shalt say,
> 'I have no pleasure in them';
> Or ever the sun and the light,
> And the moon and the stars, be darkened . . .

Because man goeth to his long home,
And the mourners go about the streets:
Or ever the silver cord be loosed,
Or the golden bowl be broken,
Or the pitcher be broken at the fountain,
Or the wheel broken at the cistern;
And the dust return to the earth as it was,
And the spirit return unto God who gave it.

During this century, however, there has come into being the phenomenon of the Third Age: we may expect to live much longer in health and independence – perhaps 20 to 30 years after retirement. Perhaps this should be regarded as beginning a new phase of life and we should, right at the beginning of it, as in adolescence, regard existential loneliness as a creative thing, and not shun it, anxiously seeking a gregarious association with our age-peers simply to avoid the spectre of a lonely old age. Perhaps we should re-examine what we really believe in and not accept unquestioningly all the conventions in which we grew up, and which have largely governed our previous lives. Society is changing in many important respects, and should we not aim to change also to meet the challenges of the Third Age?

2
The Problems of Later Life

Different forms of loneliness that may occur in later life are generally related to problems associated specifically with ageing, and in this chapter these problems will be described and discussed. The troubles and handicaps of what may be called the Third Age are of two kinds: those that are 'biogenic' and those that are 'sociogenic'.[1] By 'biogenic' we mean handicaps that arise simply through the process of becoming less robust and efficient in the course of ageing, and by 'sociogenic' we mean troubles that arise through the treatment of older people by society, and which could be lessened by more humane, sensible and unprejudiced attitudes and institutions. Social scientists reckon that the latter are far more responsible for the miseries and loneliness of later life than the former; indeed they may contribute to deterioration in physical fitness: for instance, if pensioners have to support themselves on an inadequate income they are much more likely to suffer from all sorts of physical ills arising from poor diet, inadequate housing, and a generally unhealthy life-style. Economizing on pensions may prove to be a foolish social policy if it results in greatly increasing the amount spent on the National Health Service.[2]

Biogenic problems

Human beings, like other animal species, decline in physical robustness towards the latter part of their life-span. Sight and hearing become less efficient not through any disease but simply through the natural deterioration in the ageing process; muscles weaken and joints become stiffer even if there is no actual medical disorder such as arthritis. Most of our internal organs, such as the kidneys and liver, begin to perform their functions less efficiently, and there is a general slowing of most of our responses. But the most important change with age is that our recuperative powers decline and we are less able to resist and recover from illnesses and accidents. All such changes are related to the age at which we die; thus a man of 75 years is about 41 times more likely to die during that year than a man of 20. There are, of course, very great differences between individuals as to the rate they deteriorate with ageing. The rate of decline is partly determined by genetic inheritance, and no one can do anything about the genes that are inherited, but general health is powerfully linked to this rate.

The medical profession, and all those concerned with public health, can congratulate themselves on the great progress that has been made in the present century in the matter of the health of the population, at least in the more affluent Western countries. Not only have many infective diseases such as smallpox, cholera, tuberculosis and polio been practically wiped out, but people live longer – and live their later years in relative health and independence. This is the phenomenon of the Third Age, something that has never existed in history before: quite a large segment of society have retired from fulltime work and have a span of 20–30 years of *active* life to look forward to.[3]

Some may object, 'Is it any real gain to have a large body of people who are physically fit and at leisure, but often lonely and thus constituting a mental health problem?' It is a common assumption that many older people are in fact very lonely, and rather miserable because of this loneliness, and this book examines how true this assumption is. According to various surveys 90 per cent of the population of Britain believe that loneliness is a problem for older people. In the USA a Louis Harris poll for the National Council on Ageing,[4] which was carried out in 1975 and repeated in 1981, showed that the majority of people aged 18–64 were of the opinion that people over 65 had 'very serious problems' of loneliness, poor health, not enough money to live on, and fear of crime. Most people *over* 65 thought that at least half of their own age-peers had 'very serious problems' in these areas; but none of these items were reported by the majority as *applying to themselves*. The problems from which older people suffered were often not the same as those attributed to them by the young. The gerontologist Vern Bengston commented:

> The message that emerges here is that the older public, like the young, have bought the negative image of old age.... They apparently assume that most old people are miserable and that they are merely exceptions to the rule. Myth has replaced reality.[5]

Certainly new problems have arisen because of the development of a Third Age. As already mentioned, women tend to live six or seven years longer than men, a fact that no one has ever explained satisfactorily, and hence in the later decades of life there must be more and more widows, women who may be lonely because they miss their husbands, as well as having very little chance of finding a male companion.

The relationship between physical fitness and loneliness was examined in Chapter 1, and as confirmed by the survey described in Chapter 3, there is a definite association. Although in the Third Age the majority may live in relatively good health and robustness until they 'drop off their perches', it is in the nature of age-related physical decline that it may affect only one system of the body. Thus people may remain perfectly healthy and anxious to continue their accustomed active lifestyle, but if the joints of their knees cease to function they may become practically housebound. True, with sufficient persistence – and money – one may partly overcome such a handicap with the aid of a motorized wheelchair. I know an eminent academic who continues to pursue his research work by getting around the corridors of a university library in such a motorized chair, and people will continue to visit his house because he is very stimulating company, but how many crippled people have such ability, affluence and luck? Loss of sight or hearing are rather more serious; I know another academic who is in the process of losing his sight; he fails to recognize his friends and is sometimes too proud to acknowledge the extent of his disability, mistaking the identity of those who address him. It must be a misery to him to be unable to read the books and journals that used to be so much a part of his life, and to enable him to keep up with the topics that he used to discuss with other people. The subject of deafness was also mentioned in Chapter 1, and here there is the added handicap in that it is invisible – the deaf person has no obvious prosthesis or white stick, and being perfectly fit and active in other ways appears to be merely stupid or unfriendly.

The span of years of the Third Age is so long that its members are exposed to the risks of later life for quite a considerable period, and one of these risks refers to the vulnerability of other people. The loneliness of widows has been mentioned, but it is not just the bereavement of spouses that is involved, for if we live a long time it is inevitable that we lose a lot of our old friends. We may make younger friends in later life, but that is relatively uncommon, and the bar to such friendships is that those who are younger have not shared the same world and experiences of the older friends.

A form of bereavement that is almost worse than death is when loved ones and acquaintances become demented. The seriousness of this problem should not be underestimated: Alzheimer's disease, the most common of the dementing disorders, attacks both the feeble and the robustly healthy alike.[6] It is very much related to age; its incidence is about 2 per cent in people aged between 65 and 75, but in the group over 85, the group that is becoming proportionally larger as the years go by, the incidence rises to about 20 per cent. It is not just the loneliness of the sufferers that needs to be considered, for as the disease advances they become less and less aware of their pitiful condition, but of the carers whose lives become more and more circumscribed by the demands made upon them. Carers are very often quite elderly themselves, being the spouse of the patient, and they end the marriage not only with very little life outside the home where they perform their duties as nurse, but with all the bitterness of looking after someone with whom they once had a close and loving relationship but who now hardly recognizes them, and certainly shows no gratitude for all the sacrifices that have to be made for the sufferer.[7]

The loneliness of those who look after relatives and friends who are dementing or suffering from long-lasting terminal illnesses such as cancer, has come under more widespread attention from the caring professions more recently. Recognition that it is not just the patient who needs help and support is expressed as follows in a recent textbook:

> Caring is often very stressful for families, especially in cases of dementia and when the disability exists over long periods of time. While it may not be possible to improve the elder's condition significantly, there are often modifiable aspects of the care situation. As examples, building social support and improving how caregivers manage stress can alleviate some of their emotional distress. Clinical interventions can target these modifiable dimensions to reduce stress on caregivers. Interventions can include individual counselling with the caregiver, family counselling, and support groups. In the end caregiving poses a considerable challenge to families and to society, but some straightforward clinical approaches can often be very helpful so that caregivers can assist their relative in the best possible ways without excessive burden on themselves.[8]

Some addresses which may be of assistance to caregivers are given in Appendix A.

Sociogenic problems

Many problems that cause loneliness in later life are largely the result of social attitudes to older people. The trouble is that as each generation enters the Third Age they have to contend with social attitudes towards them that are a legacy from the past. Public opinion has been changing considerably during the present century and the two great wars have been followed by considerable modifications in the lifestyles of the majority of people and what is considered to be acceptable behaviour. An outstanding example of such change has been in the area of sexual morals: in the 1930s it was still regarded as most reprehensible for women to be 'unchaste'. Premarital intercourse, although widely practised, was generally regarded as most immoral, and couples who lived together without being married would find great difficulty in securings lodgings if their unmarried state were known.

Nowadays restrictive attitudes towards sexual behaviour outside marriage persist only among an atypical minority; it is generally accepted that engaged people sleep together, and many couples live together out of marriage and are considered quite respectable. Unmarried mothers are no longer shunned by society.

However, the old attitudes persist to a great degree in relation to the behaviour of people who are considered to be 'old'. A writer to *The Lancet* comments:

> The younger generation, so liberal, so free, so uninhibited by old-fashioned conventions according to themselves, are often rigid, narrow, puritanical and censorious when it comes to the behaviour of older citizens.[9]

We may find many examples of this with regard to all sorts of behaviour on the part of older people. The young maintain the prejudices of the past concerning the grandparent generation. David Clark, a retired psychiatrist, conducted a series of seminars with elderly people who were members of the University of the Third Age, and he reported on their discussions:

> We sometimes discussed things that it was not appropriate for us to do. . . . It was certainly clear that we should not be irritable, cantankerous or drunken in public and we had the impression that sexual activity or interest on our parts was usually regarded with dismay when it was brought to the attention of the younger generation.[10]

So even sexual *interest* on the part of older people is perceived to be reprehensible by the younger generation, yet these elderly people had been young adults in the 'swinging sixties'. It is apparent that as each generation reaches a certain age-determined milestone, a ready-made set of appropriate values that they must adopt is there waiting for them, and these values do not change a great deal from generation to generation. If society has certain stereotypes about 'the old', people will expect that they must conform to these images when they reach the Third Age, and one of the traditional assumptions about later life is that loneliness is a natural concomitant of ageing.

Public assumptions about 'the old'

Alex Comfort, an eminent authority on ageing, writes as follows:

> Let us look at the stereotype of the ideal aged person as past folklore presents it. He or she is white haired, inactive, and unemployed, making no demands on anyone, least of all the family, docile in putting up with the loneliness, cons of every kind and boredom, and able to live on a pittance. He or she, although not demented, which would be a nuisance to other people, is slightly deficient in intellect, and tiresome to talk to, because folklore says that old people are weak in the head, asexual, because old people are incapable of sexual activity, and it is unseemly if they are not. He or she is unemployable, because old age is second childhood and everyone knows that the old make a mess of simple work. Some credit points can be gained by visiting or being nice to a few of these sub-human individuals, but most of them prefer their own company and the company of other aged unfortunates. Their main occupations are religion, grumbling, reminiscing and attending the funerals of friends. If sick, they need not, and should not, be actively treated, and are best stored in institutions where they can be supervised by bossy matrons who keep them clean, silent and out of sight. A few, who are amusing or active, are kept by society as pets. The rest are displaying unpardonable bad manners by continuing to live, and even on occasion by complaining of their treatment, when society has declared them unpeople and their patriotic duty is to lie down and die.[11]

Although the above is deliberately exaggerated, it contains many important truths. The label of being 'old' is an attribute known as a *master state* – possession of one unfortunate trait implies that the individual has all the other undesirable traits allegedly associated with it. Thus someone over the age of 60 may be perceived not just as an individual who happens to have been born on a certain date, but as an 'old person', and therefore presumed to be a bit slow on the uptake, and to possess the various other attributes listed by Comfort in the quoted passage – in fact 'a lonely old thing'.

This image of older people is not just a British stereotype: the

French writer Simone de Beauvoir in her book *La Vieillesse* presents her national view of 'the old' thus:

> Are the old really human? Judging by the way our society treats them, the question is open to doubt.... In order to soothe its conscience, our society's ideologists have invented a certain number of myths – myths that contradict one another, by the way – myths which induce those in the prime of life to see the aged not as fellow men but as another kind of being altogether. The Aged Man is the Venerable Sage who planes high above this mundane sphere. He is an old fool wandering in his dotage. He may be placed above our kind or below it; but in either even he is banished from it. Excluded. But what is thought an even better policy than dressing up the facts is that of taking no notice of them whatsoever – old age is a shameful secret, a forbidden subject.[12]

The above was written 30 years ago, and in the intervening years it must be admitted that in France as in Britain there has been some progress in overcoming the 'banishment' of 'the old' from society. The University of the Third Age was first inaugurated in France, and has now spread to many countries. In Britain organizations such as Age Concern have made a significant impact on the isolation of older people, and a list of such useful bodies is given in Appendix A.

The stereotype of older people as displayed in fiction and other literature will be discussed in Chapter 4, and it will be seen that loneliness is a constant theme, so everyone in a literate society has this image before them for the whole of their younger lives. It is natural therefore, that the assumptions about old age, and their influence on the self-perception of older people, will continue to persist for a very long time.

Older people in institutions

It is often assumed that a large proportion of the retired population lives in 'institutions' of some kind – 'old people's homes', geriatric wards, and for the more affluent, nursing homes. Scandals concerning badly run institutions constantly hit the headlines as being eminently newsworthy, and thus the bogey of spending one's last lonely years in the hands of uncaring strangers is held before us, just as ending life as a pauper in the workhouse was kept before the working class as an instrument of intimidation in an earlier age to make them thrifty, sober and hard-working. In actual fact only about 4 per cent of the population over the age of 65 live in any sort of institution; the rest live and die in their own homes.[13] People who are in some sort of residential care generally have something wrong with them, making it impractical to live in their own homes, and the image presented in the popular TV drama *Waiting for God* is somewhat misleading. Whether it is lonelier living in a 'home' with a group of one's age-peers, or at home coping with various disabilities, is debatable, and depends on a wide variety of circumstances, including one's personal attitude to solitude.

Retirement and disengagement

The idea that the section of the population regarded as 'old' should be sequestered from society at large in a psychological sense, although not actually relocated in a geographical 'reserve', is responsible for a certain feeling in those concerned of loneliness and of just not being wanted. Not only are they compulsorily retired from their jobs at a certain age, whether they are fit to carry on with their work or not, and irrespective of their wishes, it is implied, by various prohibitions that we will discuss, that they are not entitled to full citizenship.

It should be noted that the age at which people are retired is quite arbitrary and entirely illogical.[14] Women age less rapidly and live six or seven years longer than men on average, yet their retirement age was set at 60, whereas that for men was 65. This retirement age is currently under review, but what really needs reviewing is the whole concept of a statutory retirement age. It has been known for some time that in later life very great differences develop between individuals: some in their 80s are as young, physically and psychologically, as others in their 60s. It is simply absurd to use chronological age as a milestone. In some professions those who are officially retired are permitted to carry on with their work in a restricted sense, and even allocated an office; this is usual in the academic world. Their retirement is largely a matter of ceasing to pay them any salary. But for the ordinary manual or clerical worker retirement is as absolute as if they have been given the sack.

The concept of retirement with an old age pension really developed in Britain early in the present century and it was hailed as an important and humane development inaugurated by a Liberal government. Three main factors have been involved in the history of pensionable retirement.[15] First, demands for greater efficiency and productivity at the workplace required that firms got rid of their older workers, as it was assumed – quite erroneously, as has been shown by subsequent research studies – that older workers would be less efficient. Enlightened social policy demanded that older workers should not just be bundled into the workhouse when they ceased to earn a wage. Second, the growth of the factory system with its assembly-line system, meant that actual skill, such as is acquired by older workers of many years' experience, was no longer necessary: low-paid youngsters who had served no apprenticeship were all that was required. Third, in periods of high unemployment characteristic of the economic history of the present century, getting rid of the older section of the workforce provided a welcome relief. Their pension was generally far below what they would earn when in employment with wages determined by trades union bargaining. When we hear that the rate of unemployment is falling (and three cheers for government economic policy!) we seldom reflect that in an ageing society some portion of the fall is not due to improved economic conditions: it is *inevitable* as a larger and larger proportion of the potential work-force reach pensionable age.

If people in employment *want* to retire – fine, that is their preference, and many seek early retirement in certain occupations – but for the majority of the working population their social life depends very much upon the contacts they make at their place of work, and retirement can be very traumatic for some and lead to an unexpected loneliness, particularly in the case of men. According to Alex Comfort, who was quoted earlier:

> Most old people who are not ill do manage to handle loneliness at least as well as younger age groups by dealing with it through their own resources. One enormous reservoir of unhappiness could be drained if we expected older people to

work and allowed them to do so. Work, unless wholly solitary, is the natural antidote to loneliness.[16]

When people retire, if they are living as man and wife – and this is the norm for people reaching retirement age – the amount of time that they spend together greatly increases, although it may not be as great as one of the partners desires. This change in the lifestyle of one or both partners can have profound and far-reaching effects, and adjustments have to be made comparable to those which are necessary following marriage or early cohabitation. But when young people adjust to a new way of living, they are generally flexible in their attitudes and adapt relatively easily to a new lifestyle, learning to compromise with one another's habits; however, at the age of retirement the situation is very different. Older people may be well adapted to sharing a home, but the home may mean something rather different to the two of them. The man may regard it as a comfortable background to his 'real' work, that is, his paid occupation, and such part-time activities as home decorating, gardening, and other hobbies are generally simple relaxations which are pleasant, but for which there is never enough time.

The female partner may have comparable leisure-time pursuits which form a background to her more routine and demanding work whether or not she is gainfully employed full-time. She may have come to regard some areas or aspects of the family home as exclusively 'her' domain, and the man may have assumed that others are exclusively 'his' property. Retirement sometimes results in 'trespasses' in the other's domain, a new source of friction. Both partners may have settled into some sort of stable routine in relation to their work, leisure, social contacts and relations with each other. Most people's work and leisure habits follow some sort of routine, and while they may not provide much excitement, such patterns of habits are generally quite comfortable and peaceful as long as the existing lifestyle is preserved. Upsets may occur with retirement because the whole way of living is changed. It is generally believed that men look forward eagerly to retirement to engage more fully in various activities and projects they have never had time for, only to be disappointed when they find that they never get down to doing what they intended to do, and have time on their hands, time in which they experience real loneliness. They no longer have their associates at the work-place to foregather with, and to share their interest in sport, politics and men's gossip. Their wives are more likely to have made friends among the neighbours, and a housing estate can be a very lonely place for some people.

The reality is somewhat more complex than this and often relates to the dynamics of their marriage, for it is possible for one or both partners to be lonely within a marriage, although it may be difficult for them to admit this, even in replying to a questionnaire which is completed anonymously, as will be discussed in Chapter 3.

Even if couples have quite enjoyed one another's company in the limited hours of leisure in the working day, and on the holidays they have taken together, the new situation of being something like 24 hours a day in contact, in some cases, may place a new and intolerable strain on their emotional relationship. A new life is beginning, the life of the Third Age, and they are apt to look at and reassess both each other and themselves.

Disengagement theory

Disengagement theory was propounded by some social scientists during the 1960s.[17] It holds that it is desirable and almost inevitable that at a certain age people should distance themselves from participation and concern with the ordinary life of the community, both occupationally, socially and psychologically, and retire to a life that is more private, and more restricted. Different cultures have a variety of traditions relating to the lives of older people. In India there is a Hindu tradition of elderly people giving up all responsibility and retiring to an ascetic way of life, spending their time in contemplative pursuits and visiting sacred shrines, thus preparing themselves for life in a future world after death and reincarnation. Such a tradition applies to more affluent families where the services of older people are not needed; it does not embrace the great mass of the very poor in India, for whatever limited capacity may remain to older people can be put to some use in poor communities. Various writers have described how the old are looked after in village life, where elderly women perform tasks of child-minding and doing what-

ever simple work is within their capability. There is no disengagement there. In medieval Europe there was a tradition of more
affluent people breaking the marriage tie if they had lived to
what was then considered old age, and retiring into monasteries
and nunneries for the last years of their lives, thereby ensuring
that they would be part of a caring community rather than
facing a lonely old age, and at death they would be leading
holy lives. Naturally, in the poor peasant communities of medieval
Europe, and in some European countries today, as in India, the
grandparent generation were expected to continue working until
they died, exploited by their families perhaps, but at least not
isolated.

'The old' become a problem when they are regarded as a burden to the working population, and it is then they have to bear
the isolation and the loneliness of being made to feel unwanted.[18]
In many societies there is a pronounced ambivalence in traditional attitudes towards the grandparent generation. In Japan older
people are supposed to be entitled to special honour and respect
according to Confucian teaching, yet there is also in their tradition a cruel practice of *obasute*, which means literally 'getting
rid of Granny'. It is alleged that in ancient times unwanted old
people were simply carried up a mountain and left there to die;
how widely this was practised is unknown, but there is a folk-
song, 'The Oak Mountain Song' which describes this, and in
modern Japanese literature there is some hint of it in stories
about how families dispose of their unwanted elderly relatives.
The practice of the Eskimos in former times is referred to by de
Beauvoir:

> The Eskimo, whose resources are meagre and uncertain, per-
> suade the old to go and lie in the snow and wait for death;
> or they forget them on an ice-floe when the tribe is out fish-
> ing; or they shut them in an igloo where they die of cold.[19]

We may think that this is a far cry from what is done in modern,
civilized societies, but we may find some echo of it in current
practice. In July 1973 the British Medical Association noted that
'granny dumping' was becoming increasingly frequent. The process
involves getting an elderly relative into hospital for a spell for

some remedial complaint, and then when the patient is ready for discharge to declare that there is nowhere for her to go, and that they cannot possibly provide the necessary care or pay for the elderly person to be looked after. It is no wonder that the 'grannies' who have been dumped in this manner should feel a desperate sense of loneliness, and wonder if it is true that, as suggested above with bitter irony, 'it is their patriotic duty to lie down and die'.

Using death as a means of getting rid of the unwanted elderly population is not an idea quite foreign to civilized countries in the present century. In 1906 William Osler, the celebrated British physician, said the following in the course of a speech:

> I have two fixed ideas.... The first is the comparative use-lessness of men above 40 years of age. It is difficult to name a great and far-reaching conquest of the mind which has not been given to the world by a man on whose back the sun was still shining. The effective, moving vitalizing work of the world is done between the ages of 25 and 40 – these 15 golden years of plenty, the anabolic or constructive period in which there is always a balance in the mental bank and the credit is still good.
>
> My second fixed idea is the uselessness of men above the age of 60 years of age, and the incalculable benefit it would be in commercial, political and professional life if, as a matter of course, men stopped work at this age.... In that charming novel *The Fixed Period* Anthony Trollope discusses the practical advantage in modern life of a return to this ancient usage, and the plot hinges upon the admirable scheme of a college into which at 60 men retired for a year of contemplation before a peaceful departure through chloroform. That incalculable benefits might follow such a scheme is apparent to anyone who, like myself, is nearing the limit, and who has made a careful study of the calamities that may befall men during the seventh and eighth decades. Still more when he contemplates the many evils which they perpetuate unconsciously, and with impunity.[20]

Five years later William Osler was knighted. It is impossible to say to what degree he was speaking with his tongue in his cheek about chloroforming men over 60, but if so he was using the well-known technique of making a proposal in the guise of humour in which one believes. The same technique for the same purpose was used by Donald Gould writing in the *New Scientist* much more recently as below.

Some are questioning whether 'the old' are becoming an increasing burden on society as they are becoming more numerous. Certainly the oldest 15 per cent of the population are estimated to consume 25 per cent of the medical services; no-one likes to suggest that public euthanasia centres should be set up to reduce this perceived burden, but some responsible people like William Osler and Donald Gould have hinted at it – all in fun of course! But behind such 'funny jokes' there is perhaps a serious concern. Gould writes as follows:

> Old people don't earn any money and have to be paid a pension. They frequently suffer ills of the flesh so that their crumbling bodies clog up family doctors' waiting rooms and occupy an absurdly high proportion of costly hospital beds. Their younger relatives have better things to do than act as unpaid servants to wrinklies no longer capable of looking after themselves. In short oldies are a damned nuisance all round, and their numbers are increasing at an alarming rate. . . . There have been suggestions that before too long the state will have to impose a statutory age limit on the right to life. This could involve oldsters receiving on, say, their 75th birthday, a buff OHMS envelope instructing them to attend their local euthanasia depot on the following Wednesday at 2.30 in the afternoon.[21]

The article continues in a jocular manner suggesting ways in which 'the old' could be disposed of, and it is accompanied by a humorous cartoon to make it clear that it is not to be taken too seriously. It may be remarked that if a similar jesting article had been published on the subject of the high unemployment rate of black people in our society, one can imagine the outcry there would have been! Racism is to be condemned in all its forms, jesting and otherwise; should not ageism be similarly condemned?

Much that was written about the disengagement of older people in the sociological literature in the 1960s was, to some degree, based on a self-fulfilling prophecy. In our modern, secular society individuals have been compelled to disengage because of the social and economic pressures that were put upon them, yet this is not an entirely modern phenomenon. The fate of Shakespeare's King Lear, who felt that he should disengage in favour of his daughters, and his subsequent loneliness, insanity and death should be a warning to us all. We do not all have daughters like Goneril and Regan, but all that we know about troubles in families in this modern age points to the fact that all too often those who give up *power* in later life may face a lonely future.

Disengagement theory holds that the process of giving up rights and responsibilities at a certain age, and retiring to a metaphorical rocking-chair, is universal, and moreover, that it is to the advantage of both society and the individual. It hardly needs pointing out that a neurosurgeon cannot operate as effectively in his fifties as when he was younger, just as a coal-heaver cannot heave coal as efficiently after a certain age, but that does not mean that they should disengage either socially or economically as they age and be content to accept a lower standard of living and a less respected place in society. To the extent that they are forced to do so is an indictment of the economic and social organization of our society.

The theory has come in for a good deal of criticism in more recent years, and most specialists working in the study of ageing advise that, for the sake of both our physical and mental health, as we age we should continue to be active just as long as we can.[22] There are various meanings that disengagement may have for different individuals; thus some people may disengage psychologically but not occupationally. A journalist may continue working at his economically rewarding trade in his later years, long after he has disengaged psychologically and socially from the world of journalism, as other spheres of endeavour may have come to capture and absorb his interests. Disengagement may be most unwelcome for other people and contribute to their loneliness; it may simply reflect their diminished power, their unwilling withdrawal from interactive exchanges because social institutions such as occupational retirement have robbed them of the contacts they used to have.

It was suggested earlier in this chapter that not only were older people prohibited from continuing their work at a certain arbitrary age, but that the rights of full citizenship were denied to them. What sort of life are retired people supposed to lead? A lonely life sitting in the pub reminiscing to anyone who will listen, or sitting at home watching the television? At another stratum of society, perhaps, playing endless rounds of golf while they are still fit enough, and whiling away the time reading paperbacks?

In 1989 Age Concern sent out questionnaires to various interested groups asking about how older people should be recruited for voluntary work.[23] One question related to statutory bodies and whether there should be age-limits on appointment and retirement. Some replies stated that 'fitness' should be the only criterion, but added strange caveats such as 'provided that the member is still able to grasp and contribute to meetings'. Eric Midwinter comments on these surprising caveats:

> Why add them with their evocation of the doddering loon, slumped, with listless ear-trumpet, over the board table? Should not that proviso apply to *anybody* serving on *any* committee? Is not the question more of how we select and appoint committees in the first place?. . . .
>
> Turning now to the representatives of old age who actually believe age is an *automatic* disqualifier, let us scrutinize the sort of replies among which this was typical: 'Capability should be the main criteria (*sic*) but 70 years of age should be the limit.' Another response suggested the same limit: '70 – this allows recently retired people to use their experience but after this age their experience becomes dated.'
>
> Several quote 70 as the upper age limit for voluntary and statutory committees alike. The suggestion that experience has a 'sell-by date' is preposterous. It might be true of some forms of knowledge, but it is absolutely false in terms of experience.[24]

The point that Midwinter makes is that age-discrimination is in itself unjust and that one cannot make it any more just by

tinkering with the age-limits. It is like watering down racist discrimination against black people by allowing some the normal
rights of citizenship – but not those whose skin is *very* black, or
tinkering with sexist discrimination by allowing full rights to
most women – but not those who appear to be *very* feminine.
To practise ageist discrimination is to deny the full rights of
citizenship to some older people: it is to tell them that they are
too old to be taken seriously in the world of affairs, that they
have reached second childhood and that they should run away
and play. This denial naturally hits the more able and poten-
tially active members of the older community hardest, and to
them it is one of the cruellest of the problems of later life.[28]

3
The Measurement of Loneliness

Measuring loneliness

In order to study loneliness it is necessary to have some way of measuring how serious it is and what form it takes, for in Chapter 1 we discussed the different types of loneliness. At different periods of our lives loneliness may be experienced in different degrees of intensity, and it is not wise to accept the usual assumption that we get more lonely in later life. Obviously the *causes* of loneliness will vary greatly according to people's age: young people are often lonely because they have failed to find a mate, or they have been jilted; in later life bereavement and retirement from work are among the principal causes.

Researchers have used two main methods of gathering information: interviewing, and that of asking people to fill in questionnaires.[1] The former method is very time-consuming, and without a very great expenditure of time and money we can only deal with a relatively small sample of people, and it is therefore less certain that they truly represent the target group that is being aimed at. One of the problems in the measurement of loneliness, whatever method is used, is that some people will try to minimize their degree of loneliness out of a kind of pride – a conviction that one *shouldn't* be lonely and that it is a sign of personal failure; others may exaggerate their degree of isolation out of a general habit of self-pity, and perhaps hope that the interviewer may be in a position to alleviate their plight through the social services and by effecting introduction to clubs and organizations. The latter drawback applies particularly when face-to-face interviewing is used, but by employing questionnaires which are completed anonymously, as in the present study, people have less reason to distort their responses.

Not everyone will be prepared to complete such paper-and-pencil inquiry forms, and it may be that those who decline to participate will have characteristics different from the majority of the group studied, e.g. proud, lonely people who are very jealous of their privacy.[2] On the whole, however, researchers using well-planned questionnaires tend to get a response rate not far short of 80–90 per cent of *usable* completed returns (one must allow for those who make a mess of filling in any forms out of carelessness and misunderstanding).

Researchers have taken two different approaches to the measurement of loneliness: in the *unidimensional* approach the condition is viewed as a simple phenomenon regardless of whatever is causing the individual to be lonely. Thus a measurable degree of loneliness would apply equally to a young woman in a strange town who lacks friends and feels isolated, and an elderly woman who feels bereft because her husband has died and has few friends or family to comfort her.

The other approach is *multidimensional* and seeks to differentiate between the various manifestations of loneliness. The fact that loneliness may be found to be associated with a particular factor, say unemployment, does not necessarily mean that lack of employment *causes* people to be lonely. That would be far too simplistic an assumption. All that we can determine is that the measure of loneliness correlates positively or negatively with a particular factor, and the reason for the association is a matter of interpretation.

The present study

In the research that is reported here the intention was to study an elderly population and attempt to determine what factors were associated with greater and lesser degrees of loneliness. The population chosen were members of a branch of the University of the Third Age (known as the U3A).[3] This body is a country-wide organization of autonomous groups of people, most of whom are over the age of retirement, and its aims are partly educational and partly social and recreational. No educational qualifications are required to join, but in practice most members are fairly well educated and come from a middle-class background. The questionnaire that was used is reproduced on page 42. The questionnaires were sent out to the homes of 140 people randomly chosen from the membership list with a request that they should be completed anonymously and returned to me, men and women being equally represented. There were 121 usable questionnaires returned. In reporting the results of the analysis of the completed questionnaires a number of tables will be presented.

How the results were analysed

In the explanation and discussion of these figures, where percentages are used they are approximate, the figures being rounded to the nearest whole number. In technical books and journals statistical techniques are used in order to report how the tendencies observed differ from what may be expected by chance but, as this book is intended primarily for a lay readership, such statistical techniques will not be employed, and in many cases a commonsense inspection of the tables will be all that is necessary to appreciate what was found. Where the association between two measures (e.g. degree of loneliness and married/not married) is not so obvious, it will be necessary to express the results in quantitative terms. In order to avoid the complexities of statistical operations and 'levels of probability', a very simple index of association will be used – the Beta coefficient.[4]

Suppose we have two measures, like loneliness and marital status, and we call them X and Y, then we can construct a four-celled table like the following:

		X	
		No	Yes
Y	No	a	b
	Yes	c	d

The letters a, b, c and d represent the numbers of individuals in each cell; thus the number of people who are not lonely and

QUESTIONNAIRE

Please put a tick in just one cell in answer. Please answer each question.

	Never	Rarely	Occasionally	Often
1. Do you feel lonely?				
2. Do friends call on you?				
3. Do you try to make new friends?				
4. Do you find socializing a waste of time?				
5. You yearn for the happier days of the past?				
6. You discuss your troubles with an intimate confidant?				
7. Are you happiest in group activities?				
8. Do you find it easy to make new social acquaintances?				
9. You feel shy at parties and other social occasions?				
10. You are happy to be on your own for most of the time?				

Man.... Woman.....   *Age bracket* 60+ 65+ 70+ 75+ 80+ 85+

Marital status: Married [ ]
Widowed [ ]
Separated [ ]
Always single [ ]

Living accommodation: Living with Spouse [ ]
With Relative(s) [ ]
With Friend(s) [ ]
Communal Group [ ]
Alone [ ]

Health: Good Fair Poor
[ ]   [ ]   [ ]

Comments please
....
....
....
....
....
....

not married would be entered in cell a, and those who are both lonely and married would be in cell d. It is obvious who would be in cells b and c. If people who are both lonely and unmarried, and both married and not-lonely is comparatively large then the cells b and c would have bigger numbers in relation to the other cells, and it would be correct to say that the measures of marriage not-lonely are *positively* related.

Although the fact of this relationship would probably be obvious just by looking at the numbers in the cells, it is convenient to have some actual index of the *strength* of the relationship, so that we could, for instance, determine whether the relationship was stronger for women or for men by carrying out independent analyses. The index that is employed is known as Beta, and is obtained by the simple formula:

$$\frac{b \times c}{a \times d} = \text{Beta}$$

Whether we put (b × c) or (a × d) as the value above the line depends on which number is the larger, and the value of Beta can be positive or negative. Readers will understand this matter better when they have read through the chapter and seen some actual examples.

Age and loneliness

The questionnaire inquires about the age of the respondents, giving a range of 60–85+, in order to test the assumption that people might become more lonely as they grew older, as ageing is associated with such things as bereavement and growing infirmity. The results are shown in Table 3.1 where the younger group aged 60–70 is compared with the older group over 71 years. It is obvious from this table that there is no significant tendency for the older group to report themselves as being more lonely.

In order to investigate this matter further, analysis of the above figures was carried out for men and women separately, as various factors such as widowhood are more common in older women, but in this sample of people no such age-related tendency was found for either sex.

Table 3.1 Do you feel lonely? (Age differences)

Age	Never	Rarely	Occasionally	Often	Total
71+	17	22	18	5	62
71−	15	21	19	4	59
Total	32	43	37	9	121

Are men or women more lonely?

Table 3.2 shows the results from the 62 men compared with those from the 59 women.

Table 3.2 Do you feel lonely? (Gender differences)

	Never	Rarely	Occasionally	Often	Total
Men	24	21	13	4	62
Women	8	22	24	5	59
Total	32	43	37	9	121

It may be seen that there is some tendency for the men to report themselves as being less lonely than the women. This tendency is more obvious if the 'Never and Rarely' categories are grouped together and called 'Not lonely', and the 'Occasionally and Often' categories called 'Lonely'.

	'Not lonely'	'Lonely'
Men	43	17
Women	30	29

Beta = −2.4

Thus there is a *negative* relationship between maleness and loneliness. To put it another way, 43 (70 per cent) of the men are 'Not lonely' compared with only 30 (50 per cent) of the women, and in the 'Lonely' category the tendency is reversed. Why the women in general in this particular sample are more lonely than the men will become more apparent later.

Table 3.3 Do you feel lonely? (Marital status)

Men	Never	Rarely	Occasionally	Often	Total
Married	23	19	9	2	53
Other options	1	2	4	2	9
Total	24	21	13	4	62
Women					
Married	6	5	9	0	20
Other options	2	17	15	5	39
Total	8	22	24	5	59

Loneliness and marital status

In the questionnaire four options are given as to marital status: Married, Widowed, Separated, Always single. It should be noted that the category of 'Married' includes only those who are *currently* married; married men and women who are now bereaved or separated/divorced are not included in this category. In Table 3.3 the Married are compared with the other three options taken together, men and women being represented separately because it is well known that the marital state has rather different effects on the two sexes.

The first thing to notice about Table 3.3 is that whereas 53 (85 per cent) of the men are currently married, only 20 (34 per cent) of the women are currently married and living with their husbands. A number of factors in the Third Age account for this striking difference: women live six or seven years longer than men on average, hence there are many more women who are bereaved. Also, as men generally marry women younger than themselves, widows outnumber widowers. A further factor exaggerates this numerical difference: as there is a numerical excess of women in the upper age brackets, men have a greater chance of remarriage after they have been bereaved or divorced.

Again combining together the two sets of adjacent cells to produce a 'Not lonely'/'Lonely' dichotomy, we get the following contrast:

	Men		**Women**	
	'Not lonely'	'Lonely'	'Not lonely'	'Lonely'
Married	42	11	11	9
Other	3	6	19	20
	Beta = -7.6		Beta = -1.3	

It is apparent that there is a negative relationship between being married and 'Lonely', and that the relationship is much stronger for the male group. But if we consider the contrast between men and women for the 48 people in the unmarried group alone, males are seen to be slightly more lonely than females in the figures below.

	'Not lonely'	'Lonely'
Men not married	3	6
Women not married	19	20

It may be seen above that twice as many unmarried men are 'Lonely' (3:6), whereas the numbers of unmarried women 'Not lonely' and 'Lonely' (19:20) are approximately equal.

Widowhood, separation and spinsterhood

There are 39 women who are currently single and it is of interest to see what the condition of loneliness is in the three different states of single living as compared with the married state. The relevant figures are set out in Table 3.4.

Table 3.4   Do you feel lonely? (Single women)

	WOMEN				
	Married	*Widowed*	*Separated*	*Always single*	*Total*
'Lonely'	9	17	2	1	29
'Not Lonely'	11	8	3	8	30
Total	20	25	5	9	59

It may be seen that the largest group of women who are lonely are widows where the ratio of 'Lonely'/'Not lonely' is 17:8 as compared with 9:11 for married women and 1:8 for spinsters. No significant observations can be made about the Separated group because the numbers are too small. A number of the widows have made some interesting comments about themselves that are worth quoting. These illustrative cases will be discussed in some detail later.

Widows who are 'Lonely'

Case 39. Aged 65–70, in 'Good' health and lives on her own. She reports that she is 'Occasionally' lonely.

> *I consider myself to be a capable person, happy in my own company most of the time. However, the lack of an interesting, caring partner is keenly felt. The loss hits more keenly at certain times when one would wish to share thoughts and pleasures. I feel lonely because I have been widowed within the past year. There is now no one with whom to share everyday thoughts and pleasures. With no relatives within easy distance I feel somewhat isolated, despite having some good friends in the locality. Set times in the year are difficult and it is very necessary to make plans to be with family/friends at such times as Christmas, Easter and holiday periods.*

Case 57. Aged 65–70, lives alone, in 'Good' health but is 'Often' lonely; she writes:

> *The value of animals as companions should be considered, and the economic situation is important. A favourable family back-up eases loneliness and doing things with and for others, hobbies, eating properly, and listening to others all help. The loneliness of wanting to say 'We' instead of 'I'. My own company can become boring.*

Case 59. Aged 75–80, in only 'Fair' health and 'Often' lonely. This lady lives in a small village some distance from the nearest town, which is perhaps significant in her case. She writes:

> *The root of the problem is 'Sans eyes, sans teeth, sans taste, sans everything!' And since I wear trousers a lot of the time I suppose I could say that I don't enjoy being a 'Lean and slippered pantaloon', though a lot of people are very kind to me. I would suggest a question on 'Why?' or 'For what are you lonely?' In my case it*

is a lack of intelligent conversation. The subjects on which I would enjoy it are pretty wide-spread, but in this small community anything beyond 'chat' is found intimidating, and 'chat' bores me after a bit.

Widows who are not lonely.

There were eight widows who reported that they were 'Rarely' lonely; it is perhaps significant that no widows reported that they were 'Never' lonely although eight other women did.

Case 2. Aged 65–70, living on her own, in 'Good' health. She adds a comment that illustrates the old saying that 'time heals' but does not indicate just what has happened during the 13 years since she was bereaved. In answer to Question 6 on the questionnaire she indicates that now she 'Never' yearns for happier day in the past, i.e. before her husband died.

I have answered as my feelings are now but I would have felt differently after my husband died 13 years ago.

Case 24. Aged 80–85, living on her own, in 'Good' health. This lady illustrates a very conventional case of a widow who is lucky enough to be still enmeshed in a family yet lives on her own. She appears to live very much within the family for on the questionnaire she indicates that she 'Often' finds socializing a waste of time, she 'Often' feels shy at parties and social occasions and 'Never' finds it easy to make new social acquaintances.

I rarely feel lonely because I have a close supportive family and grandchildren.

Case 45. Aged 73, in only 'Fair' health, but has formed a close and meaningful relationship with a man.

I am not the usual 73 year old widow living alone as I have a widower neighbour friend with whom I spend much time. If it were not for him my answers to loneliness questions would be very different. What I remember of widowhood experience before I moved here and met my widower friend was that I was utterly dismal and lonely and that family and friends only partly filled the void. I was given the advice never to refuse an invitation for a whole year, but I think the best therapy is keeping busy.

Women who have never married

There are nine women who have never married and only one is in the 'Lonely' category. All of them live alone, except for one who lives with a friend. As may be seen from Table 3.4, these spinsters report themselves as being, on the whole, less lonely than the married women; one explanation of this may be that in the later years of life some elderly women are living with husbands even older than themselves and who may not be very stimulating company, and indeed some may be frail and need a lot of looking after. The spinsters have no such duty that may tie them to the house a lot of the time; they have had a lifetime of adapting to a single state as far as marriage is concerned, and to living on their own.

Case 27. Aged 70–75, in 'Good' health and 'Rarely' lonely. She is of a scholarly disposition and is obviously only too happy to be on her own for most of the time. She writes:

I am a biblioholic so the greatest problem is stopping reading and studying. . . . I prefer my own company. I use the phone to contact friends and relatives. I like living alone but I also like visiting friends. My deepest dread is the potential death of close friends whose contact I greatly value.

Case 31. Aged 65–70, in 'Good' health and 'Rarely' lonely. This case highlights what many women mention – that in later life some people live on the brink of the loneliness which may follow some accident that restricts mobility. She writes:

I am sure being on my own for many years has made it much easier to cope with the odd patches of loneliness in later life. I usually enjoy good health but a recent broken ankle has shown me how easy it is to be lonely with restricted mobility, or pain/tiredness etc.

Table 3.5 Do you feel lonely? (The marital status of men)

	Never	Rarely	Occasionally	Often	Total
Married	23	19	9	2	53
Widowed	0	1	2	1	4
Separated	1	1	1	1	4
Always single	0	0	1	0	1
Total	24	21	13	4	62

Case 40. Aged 60–65, in 'Good' health and 'Rarely' lonely. She complains that she is partially deaf and that this affects her ability to make new acquaintances or to participate in social occasions. She writes of her rare occasions of loneliness:

> *My loneliness is exacerbated in groups owing to my partial deafness – it is not so easy to establish new friendships. . . . Physical distance from friends means that the telephone is a vital link for me and my friends as I have few friends who live nearby.*

The women who are separated/divorced

There are only five of these cases and they all live alone; as might be expected they are not as well adjusted to the single state as those who have never been married. The questionnaire does not ask how long they have been separated, and as a group they are quite reticent about themselves.

Men who are currently not married

Out of the 62 men there are only nine (15 per cent) who are not married currently, and they all live alone. In the previous analysis it was shown that as a group these men were rather more lonely than the unmarried women. The figures relating to the marital status of the men are given in Table 3.5.

It may be observed in Table 3.5 that those who are not married are scattered all over the cells, a distribution that indicates that in the unmarried state a wide variety of factors account for their degree of loneliness, but with such a small number involved we cannot make any definite statement. The following cases are of interest:

Case 129. He is the only man who has always been single and he makes the comment, 'Never been bored or lonely in my life!' In replies to the questionnaire he indicates that he 'Often' finds socializing a waste of time; he 'Rarely' tries to make new friends; he 'Often' feels shy at social gatherings, and is happy to be on his own for most of the time.

Case 118. He is separated and 'Often' feels lonely, yearning for the happy time in the past, and friends 'Rarely' call on him. He gives the following information:

> *The past time I yearn for mostly is only 14 years ago, when I lived with my young friend Joan. . . . Shortage of someone with whom to visit shows, museums, meetings, or holidays etc. I often feel that it would be nice, if I live on my own, to be within easy visiting distance of friends or family where I could drop in occasionally. But at the same time I don't want to live in my home town and be smothered by my extended family.*

This man is aged 60–65 years and he refers to the woman he used to live with as a 'Young friend'; such a case illustrates one of the drawbacks of an 'age-gap relationship' – that the younger partner will desert and leave the older one lonely. In this case even after a lapse of 14 years he found no one to replace her.

Married men who are lonely

Although the married state protects most men from being lonely, Table 3.5 shows that there are 11 married men who are 'Often' or 'Occasionally' lonely.

Case 139. Aged 70–75, in 'Poor' health and 'Often' lonely. Friends 'Rarely' call on him although he 'Never' finds socializing a waste of time and 'Never' feels shy at social gatherings. Although he has a wife, he 'Never' discusses his troubles with an intimate confidant. He suffers from high blood pressure and arthritis, and has had an ileostomy. As well as these physical ills he has many longstanding troubles of a personal nature which he describes:

*Intellectually deprived in childhood. . . . my sister was 'the' favourite
child in our family. Thus a 'numbness' inside from boyhood, means
that there seems a permanent 'sameness' of life difficulties . . . rather
than ups and downs of happiness/unhappiness.*

This case illustrates very well what has been previously identified as 'emotional loneliness' which in this unfortunate man's
case is exacerbated by his many physical ills.

Case 144. Aged 60–65, in 'Good' health and 'Often' lonely. Friends
'Never' call on this man and he 'Often' tries to make new friends,
'Often' being happiest in group activities. He appears to attribute
some of his lonely state to the condition of his wife:

*My wife has a longstanding obsessional illness so that the house
is never tidy enough to entertain (or clean enough). My last friend
came 17 years ago. Last dinner party 24 years ago. Both of us
much better after prolonged psychotherapy.*

Case 153. Aged 60–65 years, in only 'Fair' health and 'Occasionally' lonely. He seems very conscious of growing old. He 'Often'
discusses his troubles with an intimate confidant who is presumably his wife and presents a fairly normal pattern of answers
to the questionnaire. His condition of having thyroid trouble,
which is of course remediable, may account for his less than
optimal social state. He writes:

*I find that as I get older I am getting more lazy socially – I just
can't be bothered. However, I have recently been diagnosed as having an underactive thyroid gland – which may have something to
do with it*

Case 155. Aged 65+, in 'Poor' health and 'Occasionally' lonely.
The pattern of his answers to the questionnaire appears to be
very like that of Case 153 above; like him he discusses his troubles
with an intimate confidant (presumably his wife) and this is one
of the major benefits of marriage in later life. He writes:

*I suffer from a neurological illness resulting in deteriorating mental
and physical ability which affects my social life.*

Health

We would expect, on grounds of common sense, that poor health would be positively related to loneliness in later life, as those who are in indifferent health cannot get out and about as much as those who are healthy. Again, those who are semi-invalid tend to be preoccupied with their own troubles and less interested in social affairs. Cause and effect interact: while poor health may generate loneliness there is plenty of evidence which shows that loneliness often undermines health.[5]

The questionnaire gives three options regarding Health – 'Good', 'Fair' and 'Poor'. If one doubts that such self-rating is really a good measure of how healthy people are, consider the following quoted from the Report of the British Health and Lifestyle Survey:

Self assessment of health as excellent, good, fair or poor . . . may be thought to be a measure of very doubtful validity. There is evidence, however, that in fact it correlates very well with health assessed in more objective ways.[6]

One of the findings of the present investigation that seems surprising superficially is that out of 121 men and women aged 60–86 only three people reported themselves as being in 'Poor' health. This may partly be due to the fact that it is a self-selected sample, for many people who are in really poor health do not have the energy and initiative to join a branch of the University of the Third Age (U3A) and benefit from its educational and social amenities. However, the records of this branch of the University show that there is a steady death toll of active members, the men dying off at three times the rate of the women, so in objective terms there must be many more people in poor health in this sample than their self-report implies. However, objective and subjective measures of health have been shown to differ greatly, and strangely enough, it has been found in scientific surveys that the *subjective* report of health status is a better predictor of longevity than an *objective* medically established report.[7]

What appears to happen with regard to people's report of their health status is that as they grow old they are continually altering their criteria so unless they are in a quite considerable state of distress and disability they continue to report their health as being 'Good' in their eighties even though it is quite obviously much worse than what it was in their sixties.[8]

The figures relating to health are shown in Table 3.6 below, 'Poor' and 'Fair' having been amalgamated, and the data represented separately for men and women, as the health of men has been shown to deteriorate more rapidly in later life than that of women.

If we combine 'Never' and 'Rarely' together as 'Not lonely' and 'Occasionally' and 'Often' as 'Lonely', then the following figures may be derived from Table 3.6:

		Men		Women	
		'Not lonely'	'Lonely'	'Not lonely	'Lonely'
	Good	48	10	22	19
Health					
	Fair/Poor	7	7	8	10
		Beta = −4.8		Beta = −1.4	

Thus 'Good' health is negatively related to 'Loneliness', and the relationship is stronger for men. The reason for this sex-related difference is unclear. The fact that 18 (31 per cent) of the women rate themselves subjectively as being in 'Fair'/'Poor' health, as compared with only 14 (23 per cent) of the men, is to be expected, since medical statistics reflect a greater tendency of women to report more ill-health.[9]

Table 3.6 Do you feel lonely? (Health status)

Men Health status	Never	Rarely	Occasionally	Often	Total
Good	21	17	8	2	48
Fair/Poor	3	4	5	2	14
Total	24	21	13	4	62
Women					
Good	7	15	17	2	41
Fair/Poor	1	7	7	3	18
	8	22	24	5	59

Table 3.7 Reasons for loneliness (total replies: 17 men, 26 women)

	Frequency endorsement %	
	Men	Women
A. Loneliness which represents a desire to continue a relationship with someone no longer available.	24	62
B. Loneliness which derives from no longer feeling that one is the object of love.	18	19
C. Loneliness due to absence of anyone you care for.	41	35
D. Loneliness for a personal relationship similar in depth to a lost relationship.	24	35
E. Loneliness for want of someone within the dwelling unit.	18	12
F. Loneliness due to no one to share your work or activities with.	18	23
G. Loneliness for a former lifestyle and its activities.	47	38
H. Loneliness due to a drop in status.	18	4
I. Loneliness resulting from the loss of someone who helped you form relationships with other people.	0	8
J. Loneliness due to your own lack of social skills and ability to make new friends.	24	15

Reasons for being lonely

In addition to the questionnaire reproduced earlier being sent to the sample of 140 members of the U3A branch, a form listing ten possible reasons for feeling lonely was included, with a request that respondents should indicate which items applied to them. They were also asked to put the items they chose into rank order of importance. There were 17 completed forms returned by men, and 26 by women, and the analysis of the results is shown in Table 3.7.

While very few individuals ticked only one or two items, quite a fair number ticked up to seven, so to make an analysis meaningful only the first three items in terms of their rank order were considered.

Comments on Table 3.7

There are many meaningful differences as to how men and women have responded to the various items in Table 3.7 and it is worthwhile dealing with each individually.

A. Of all items this was endorsed most frequently by the women, and it relates most obviously to the greater prevalence of widowhood in later life. It also may relate to the fact that, in some cases, women miss their grown-up children more than men do.

B. As the percentages endorsing this item are small, the difference between men and women is of little significance. Although the married state has been shown to be very protective of loneliness for men, a study involving this number of people must include marriages that are not supportive. Two married men and one married woman endorsed this item, indicating that their loneliness derived from no longer feeling that they were the object of love.

C. Here there is no significant difference between the percentages endorsed by men and women. Again two married men and one married woman endorsed it.

D, E and F. Little need be said about these items and there is no significant difference in the responses of men and women.

G. This item was endorsed by the men more frequently than any of the other items, and they gave it the highest rank order, which was not the case with women. It appears that for men after the age of retirement it may be that a change in lifestyle is much more important than it is with women as a potential source of loneliness.

H. Although agreement with this item concerning loneliness following a drop in status was infrequent, it appears to concern men rather than women. It was accepted by just one woman, and she was separated; she also endorsed F (referring to absence of companions in work and leisure) and J (referring to her own lack of ability to make friends). Probably the status of some middle-class men who have had responsible jobs may be very important to them.

I. Only two people, widows, endorsed this item.

J. This refers to one's own lack of social skills. For one man it was the only item which he held responsible for his loneliness, and three others referred also to their change in lifestyle. The four women who endorsed it had little in common otherwise.

Additional reasons for feeling lonely

The form sent out invited people to add their own reasons for feeling lonely additional to those which were suggested. The following were among those supplied by the respondents.

Women

Case 29. Married, aged 70–75, 'Good' health and 'Occasionally' lonely.

> *I would like to stress that intellectual loneliness is sometimes difficult to overcome. The loneliness of old-age deafness is devastating – no music, theatre, and quite often even conversation; it affects both partners. There must be other age-related conditions which affect the person with a feeling of loneliness.*

Case 41. Separated, aged 65–70, 'Good' health, 'Occasionally' lonely.

> *As people get older they are less willing to welcome new friends.*

Case 44. Separated, aged 70–75, 'Good' health, 'Often' lonely.

> *Occasional changes in society have the effect of making me feel marginalized, and I know that this feeling will increase as time passes.*

Case 47. Widowed, age 75–80, 'Good' health, 'Rarely' lonely.

> *The main causes due to lack of conversation in this small community and the increasing inability to travel to see friends where it would be available. Add to this the number of friends who have died and are therefore not there to telephone or write to.*

Men

Case 140. Married, aged 65–70, 'Good' health, 'Occasionally' lonely.

Having had a full and rewarding life – but now as a grandfather I feel that I am well and truly in God's waiting room – this sometimes makes me feel sad and a little lonely. Also, shop assistants are becoming very impatient with me!

Case 149. Married, 80–85, 'Poor' health, 'Rarely' lonely.

Loneliness because of a physical defect, e.g. deafness or blindness; because of removal from a loved house, garden or countryside; because of being cut off from treasured intellectual, sporting, or artistic activities; because of loss of a pet.

Case 159. Married, aged 60–65, 'Good' health, 'Rarely' lonely.

Loneliness can come when one is depressed.

Conclusions

It must be emphasized that this study was carried out with a sample of elderly people who were predominantly middle-class, and living in a fairly affluent area. Any conclusions that may be drawn from it do not necessarily apply to elderly people of a different class background and living in a dissimilar area. One drawback is that only 86 per cent of those approached returned adequately completed questionnaires, so the non-respondents may have had characteristics different from those who replied, but this is a source of error which applies to all surveys of this kind.

Referring back to the beginning of this chapter, where two main approaches to the measurement of loneliness were mentioned, it will be seen that we have used the *multidimensional* approach, and a fair number of different factors leading to loneliness have emerged, as shown in Table 3.7 and those volunteered by some of the respondents. The principal finding of the study is that living with a spouse is the main protection against loneliness in later life, and this applies particularly to men. The experience of loneliness, and the factors engendering it, appear to be rather different for men and for women. The great numerical disparity between men and women in the Third Age means that women are likely to be more lonely, on the whole, than men.

Here we must admit a further possible source of error: due to the modern change in attitudes towards the respectability of sexual relationships outside marriage, a number of those who are widowed, separated or always single, may be involved in extramarital relationships, and although living separately officially, spend as much or more time in one another's company as some married couples, and rendering one another moral support.[10] The 73-year-old widow (Case 45) whose statement has been quoted is typical of those who live alone officially but has a close relationship with a man which has made all the difference to her present life, and she compares it with her lonely state before she met him.

A caveat must be made about the interpretation of the figures under the heading of 'Reasons for loneliness'. What is reported is the factors to which individuals *attribute* their loneliness, but they may not always be correct in their attribution. Thus a widowed man may blame the death of his wife for his present state of loneliness, but he might have been just as lonely in the years before his wife died, and his present feeling of loneliness may be principally due to his not being able to come to terms with the changed nature of society, so that he feels strangely bereft and like a fish out of water without appreciating the principal cause of it or the remedies that he could seek to obtain. The difficult matter of people not being sure as to how to attribute their loneliness is a question that has received much attention from social scientists as in the following passage:

> In arriving at the conclusion 'I am lonely', people use affective, behavioral and cognitive cues. The affective signs of loneliness are often diffuse. Loneliness is a distressing emotional experience; severely lonely people are profoundly unhappy. It is unlikely, however, that affective cues alone are sufficient to identify an affective experience as loneliness. There is no unique set of emotions associated with loneliness. . . . Although the experience of negative affect may alert people that 'something is wrong' in their life, it will not invariably lead to a self-diagnosis of being lonely, rather than being depressed, overworked or physically ill.[11]

The answers to the ten questions that appear in the original questionnaire have not been fully analysed here. This would be a major undertaking involving considerable statistical analysis, and it has not been considered necessary for the purpose of this book. What has been done is to use some of the answers to add to the information that has been quoted with respect to certain individuals who have been referred to as illustrative cases.

4
Loneliness in Literature

The subject of loneliness in literature is so vast that it is difficult to know how to approach it within the limits of one chapter. It has been a theme treated so often throughout the ages that all that can be done here is to call attention to its use in different periods, and to see how society has regarded loneliness according to the changing *Zeitgeist*. Sometimes writers have concentrated on loneliness associated with old age, as in the Book of Ecclesiastes, which has been mentioned in a previous chapter, and in more modern times loneliness is treated as a problem throughout adult life. Occasionally the loneliness of childhood is dealt with, as in a number of Charles Dickens's books which refer to his own childhood. Mark Twain's Huck Finn said, 'I felt so lonesome I most wished I was dead' and his remedy for his lonely fits was to fall asleep; Holden Caulfield in J.D. Salinger's *The Catcher in the Rye* experienced it more or less as a permanent problem, and he too regarded death as the final solution for loneliness: 'I felt so lonesome, all of a sudden, I almost wished I was dead.' The present book is about loneliness in later life, but it is important to realize that the things that make people lonely are really the same in childhood, mid-adulthood and old age. We are really the same people in our later life, although we have to cope with new problems that come with our age.

Some readers may be surprised at my choice of the few examples of loneliness in literature, for I mention only in passing, or omit entirely, many important authors who have made significant contributions to the literature of loneliness and alienation:

Cervantes (*Don Quixote*), Thomas De Quincey (*Diary of an Opium Eater*), George Eliot (*Silas Marner*), Emily Brontë (*Wuthering Heights*), Dostoevsky (*Crime and Punishment, The Idiot*), Thomas Hardy (*Jude the Obscure, The Mayor of Casterbridge*), Herman Melville (*Moby Dick*), Jack London (*Martin Eden*), Thomas Wolfe (*Look Homeward Angel*), W.S. Maugham (*Of Human Bondage*), Thomas Mann (*Death in Venice*), Franz Kafka (*The Trial*), James Joyce (*Ulysses*), Marcel Proust (*Swann's Way*), Ernest Hemingway (*The Old Man of the Sea*), William Golding (*Pincher Martin*). I do not discuss the important contributions of the existentialist writers such as Sartre, De Beauvoir and Camus who have had considerable influence on other authors in this field, nor do I discuss modern playwrights, or poets such as S.T. Coleridge, W.B. Yeats, T.S. Eliot and Edward Lear although in poetry loneliness and the emotions it awakens can be evoked most poignantly. Many who have experienced periods of acute loneliness may well have dwelt on the moving verses of Edward Lear, although what he wrote was supposed to be 'nonsense verse'.

What I have tried to do in my limited selection is to choose examples of authors who have vividly evoked the spirit of their time in relation to loneliness, and whose work, more recently at least, will be well known to the general reading public.

The medieval period

In the Middle Ages society in Europe was dominated by ideas that stemmed from the Bible, and naturally Man's relationship with God and the philosophical problems of his eventual death were the theme that was dealt with in most art and literature. Literature as we know it burgeoned in the fifteenth century with the invention of printing in the West, but as the mass of the people remained illiterate for centuries to come, this literature had to be conveyed to them by the spoken word and by dramatic performances. The Mass itself was, in a sense, a dramatic performance, the re-enactment of the Last Supper, and from the twelfth to the fifteenth century the Mystery plays portrayed the main events of the Christian legends to the common people in a very simplified form. In the fifteenth century there arose a new and more sophisticated way of conveying religious ideas to the illiterate, and via the printed word, to those who could read: the morality plays.[1]

While the Mystery plays were simple enactments of supposed events from the past, the morality plays relied on allegory to convey their message. One of the best known morality plays that has come down to us is *The Summoning of Everyman* which, as its title indicates, dealt with the eventual summoning that everyone expected to experience at the end of their lives to the presence of their Maker to give account of themselves.[2]

The Summoning of Everyman

In this dramatic performance actors take the parts not only of actual persons – such as Everyman, Felawship (his familiar friend), Cosyn (his cousin), but attributes such as strength, beauty and discretion. One of the important protagonists in the first part of the play is Goodes, who represents Everyman's love of his material possessions, and in the second part the most important person is Good Dedes who proves to be Everyman's eventual saviour who accompanies him to the grave.

When Everyman receives his summons from Death, his journey to meet his Maker is represented as a pilgrimage, and he asks his familiar friend Felawship to go with him on this journey. At first Felawship agrees to bear him company, but when he hears the exact nature of the pilgrimage, he cries off. Everyman then calls upon different members of his kindred to accompany him but they all refuse, leaving him to proceed alone, a prospect that terrifies him. Seeking comfort in his loneliness he next turns to Goodes, the actor who personifies his love of material wealth, to go with him and thus allay his solitude, but the latter refuses and reminds Everyman that he would make a very dubious companion, for love of riches may result in his damnation.

This constitutes the first part of the play, and is followed by a long soliloquy in which Everyman expresses his grief that he has been deserted by his friends and family and is left forlorn, thus facing the reality of life – that we are all essentially alone. He muses on his sinfulness and repents, but recalls that at least he has done some good deeds in his life and entreats Good Dedes to accompany him on his lonely pilgrimage. Thus begins the second part of the play.

When called upon, Good Dedes responds that she would will-
ingly accompany Everyman but that she has been so grossly
neglected for so long that she is too weak to stir:

> Here I lye, colde in the grounde.
> Thy synnes have me so sore bounde
> That I can not stere.

Everyman then reflects that as all external things have failed
him he must call upon his own resources, and severally calls
upon Knowledge, Confessyon, Beaute, Strengthe, Dyscrecion, etc.
including attributes that would mean little to us today but were
important in the medieval view of Man. The action of the play
is long and contains theological arguments that were supposed
to convey to the common man some understanding of the teaching
of the Church of the time. Everyman is shriven and granted
absolution, and his act of contrition gives strength to Good Dedes
who revives and becomes his companion:

> I thanke God, noe I can walke and go,
> And am delyvered of my sykeness and wo.
> Therefore with Everyman I wyll go and not spare;
> His good workes wyll I helpe hym to declare.

Later, Everyman has to be reconciled to the fact that at the
end of his life he must be stripped of the company of *everything*,
including beauty, strength, discretion and all that makes him human;
only Good Dedes, the merit he has acquired by virtuous acts,
remains with him, and he is welcomed at the grave by an angel.

The above is a very simplified account of a quite complex play.
There is a great deal of scholarly literature about *Everyman* and
some controversy as to whether it is largely a translation from a
Dutch morality play *Elckerlijc*.[3] We may wonder today why
Everyman is represented as being so frightened of proceeding
alone on his pilgrimage, whereas other pilgrims, like Bunyan's
Christian, have accepted the loneliness of such a pilgrimage with
fortitude. It may have been that the loneliness that came from
total rejection by one's fellow men appeared to the medieval
mind as equivalent to the extinction of one's personality. An

echo of this may perhaps be seen in the final scene of Ibsen's *Peer Gynt* where Peer at the end of his life finds himself totally alone and fears that he will be melted down in the Button Moulder's ladle, along with other warped personalities. This matter will be discussed later.

It has been pointed out that during the medieval period there was a general horror of performing a long journey alone, and this antipathy to solitary travelling persists in some cultures. In modern Japan there is still some vestige of this attitude, at least among the older generation: the solitary traveller is regarded with compassion and perhaps suspicion as though he were the victim of some misfortune and an eccentric.

Today we make a distinction between loneliness and the solitary state, and in some circumstances we may regard the latter as pleasant and desirable. In the Middle Ages solitude was acceptable only in a strictly religious context; the lay person was supposed to be a 'political being'; that is, he lived, worked and travelled in association with his fellows. St Thomas Aquinas, the thirteenth-century theologian, referring back to the philosopher Aristotle, stated:

> If a man should be such that he is not a political being by nature, he is either wicked – as when this happens through the corruption of human nature – or he is better than man – in that he has a nature more perfect than that of other men in general, so that he is able to be sufficient to himself without the society of men, as were John the Baptist and St Anthony the hermit.[4]

The wandering knights in Malory's romances were solitary because they were travelling on a dedicated mission, and were thus acceptable.

With the Renaissance, attitudes began to change. Petrarch's *De Vita Solitaria*,[5] which was widely read by educated people in the fourteenth century, dealt with the pleasures of solitude in secular life. In his own day Petrarch was considered very eccentric, but gradually the old theological outlook broke down and this was reflected in the writings of the innovative Elizabethan authors who were influenced by Petrarch and similar Renaissance writers.

Shakespeare

Shakespeare's plays and verse often deal with both solitude that is unpleasant – what we would call loneliness – and solitude that is deliberately sought and enjoyed. The main works that deal with loneliness are tragedies: *Hamlet, Richard II, Richard III, Macbeth, Coriolanus, Timon of Athens* and *King Lear*. In *As You Like It*, which is one of Shakespeare's most important comedies, the theme of solitude runs right through the play.

In the Sonnets and many of the comedies, the question of solitude is debated with insight and wit. Although the nature and meaning of the Sonnets has been widely debated, it is evident that in the earlier verses the young man who is addressed is being exhorted to abandon his all-too-evident pose of the new, 'Renaissance Man' who is sufficient unto himself, and to give more consideration to his fellows and to posterity. *As You Like It* is best understood by considering it in relation to *Rosalynde*, the prose romance by Thomas Lodge from which it derives. The latter was a conventional 'pastoral', a genre of romances that were written in a number of conventions. The 'pastoral of happiness' dealt with the loves of idealized shepherds and shepherdesses, and this Shakespeare introduces in the persons of Silvius and Phebe, and mocks in the down-to-earth countryfolk William and Audrey.

There is also another genre of pastoral, the pastoral of solitude, and this Shakespeare introduces and debates in his comedy. The element of solitude is first introduced in the sub-plot of the Duke who has been banished to the Forest of Arden by his wicked brother Frederick. The Duke lives with his retainers in philosophic retirement from the court as related in their song:

> Who doth ambition shun,
> And loves to live 'i th' sun,

and so on.

In a sense, Shakespeare's play is a satire on the conventional pastoral. He introduces the solitary and cynical figure of Jaques who mocks at the alleged contentment of the Duke and his followers, and also shows his contempt for the accepted code of fellowship and good manners.

Jaques. I thank you for your company; but, good faith, I had
as lief have been myself alone.

Orlando. And so had I; yet for fashion sake I thank you too
for your society.

Jaques. God b' wi' you! Let's meet as little as we can.

Orlando. I do desire we may be better strangers.[6]

The 'melancholy Jaques' has come to be regarded as the epitome
of the malcontent who rejects society and makes a public show
of his lonely individuality. Another loner in the play is Touch-
stone, the professional fool who is equally as cynical as Jaques,
but covers his barbed witticisms with a veneer of folly. Not for
him the prospect of a wife and home; in order to sleep with
Audrey he is prepared to go through a bogus form of marriage
offered by a dubious priest, but admits:

I were better to be married of him than of another: for he is
not like to marry me well; and not being well married, it will
be a good excuse for me hereafter to leave my wife.[7]

While Jaques' condition may be one of elective loneliness, Touch-
stone is presented as one quite content with his solitary state
outside society and mocking at it. In other plays also Shakespeare
introduces court jesters who are cynical loners who mock at the
accepted social norms; Feste in *Twelfth Night* is a prime example.
It has been suggested that in real life these men were often un-
frocked priests, intelligent and learned, but unwilling to come
to terms with society.

The figure of King Lear is often presented as the epitome of
the lonely old man, abused and neglected by his ungrateful family,
but the play of that name is much more than that. Shakespeare's
repeated theme of the ingratitude of many who have been the
recipient of great favours occurs again, as in *Timon of Athens*,
but *Lear* is a highly complex play which contains also the
echo-plot of the bastard Edmund's treachery to his father and
half-brother. Lear's lonely state is shared by a number of other
characters: Kent, Edgar, Gloucester and the Fool, all of them
victims of injustice; and it appears that the play is about those
who are wronged suffering a lonely state, until – for some of

the characters – amends are made right at the end.

However, there is an alternative view of the play which links it with the medieval morality play *Everyman* which has been discussed above.[8] Although most of Shakespeare's work is very definitely post-Renaissance, there are still strong links with the past. Lear is by no means an innocent victim; like Everyman he was concerned with a love of riches and power. He regarded it as automatic that, in his position, kith and kin should support him irrespective of his deserts. He wishes to give up all the responsibility of kingship yet retain the honour and power that went with it. He is much gratified by the absurd deference shown to him initially by Regan and Goneril, and when Cordelia speaks to him dutifully but plainly he is furious and disowns her. Similarly when the Earl of Kent questions his treatment of Cordelia he loses his temper and pronounces a sentence of banishment on him.

We may regard the play as Lear's pilgrimage towards his grave in the medieval sense, becoming slowly detached from his worldly vanity, pride and abuse of power. The parallel between *King Lear* and *The Summoning of Everyman* is too obvious to miss, and here Shakespeare is using loneliness as a device for the purgation of a sinful man, and if there is no angel to greet him at the end, at least there is his loving daughter Cordelia who forgives him, and he reaches a peace of mind in repentance:

> Come, let's away to prison.
> We two alone will sing like birds i' th' cage;
> When thou dost ask me blessing, I'll kneel down
> And ask of thee forgiveness; so we'll live. . . .[9]

In *Hamlet* Shakespeare opens a new chapter in the literature of loneliness. Certainly there had been other portrayals of self-examining loners who addressed the audience in introspective soliloquies at that period, such as Richard III, and in Marlowe's *Jew of Malta* and *Dr Faustus*, but these essays in self-examination were to explain their villainy. Hamlet was not a villain, nor was he in any way deformed or warped; he was not a burnt-out rake like Jacques parading his loneliness publicly as a pose. He is presented as having been a perfectly ordinary and sociable young man until he became changed by the death of his father and the ghost's injunction that he should avenge his murder. Like Orestes he is caught in a dilemma: to bear the guilt of failing to avenge his dead father, or to incur the guilt of murder.

The psychoanalyst Ernest Jones made great play with the fact that Hamlet's delay in taking revenge was bound up with the fact that his mother had married his father's killer – a predictably Freudian interpretion of the dilemma seen in 'Oedipal' terms – Hamlet suffering from sexual jealousy, and having wished to murder his father himself.[10] Be that as it may, Hamlet's lonely musings were something new in literature. His alienation from society was not because he himself was an ambitious villain like Macbeth, or otherwise warped, as pointed out above, but because he had come to see society as rotten, and to preserve his integrity he had to stand outside it. He adopted various masks, including the mask of feigned insanity, not to deceive honest people for dishonest ends like Richard III, but to avoid seeming to connive at the false show of the Court and pretend that all was well. His over-scrupulous musings that inhibited him from action were only overcome when he broke out in extreme anger – in his mother's bedchamber, murdering Polonius, mistaking him for the King, and when finally dying from the envenomed sword, having seen his mother die before his eyes. There is some echo of this in Sartre's play *Les Mains Sales*[11] where a lonely young man cannot carry out the political assassination he intends because of his philosophical scruples, yet promptly murders his victim when a highly personal motive prompts him.

Like the existentialist writers of the twentieth century, Hamlet's portrayal of loneliness poses more problems than it answers. Had Shakespeare anticipated the existential movement before his time?

Defoe and Swift

These two eighteenth-century writers made a new contribution to the literature of loneliness in their different ways. The difference in their message has been the subject of a great deal of comment by critics throughout more than 200 years.[12] Daniel Defoe wrote for a readership of people such as himself – middle-class merchants and tradesmen, who formed a growing element in the society of the time, and some of whom were expressing themselves in the Dissenting Christian movement. Jonathan Swift, on the other hand, appealed mainly to the country gentry and those who took a Tory stance resisting innovative change. However, the similarities between them are very obvious in that both were writing at a time when the colonial expansion of western Europe of the previous two centuries had raised new problems for the civilized world.

The two works that will be examined here, exemplifying their discussion of loneliness, are *Robinson Crusoe* and *Gulliver's Travels*. The worldwide explorations of the maritime powers had produced a huge literature concerning 'castaways': individuals and groups of civilized people who had been abandoned on uninhabited islands by shipwrecks and other naval mishaps. How did civilized man react when cut off from normal society? Did he revert to being brute in his loneliness, or did he, in solitude, benefit from being removed from the corruptions of society and form a better relationship with God? What were his relations with 'savage' peoples? Should he set out to civilize them and bring them the benefits of the Christian religion, or were such efforts mere humbug to excuse the exploitation and enslavement of innocent and simple people and robbery of their lands? These questions had been endlessly debated by writers of the fifteenth and sixteenth centuries and Defoe and Swift generally took opposing views.

It is well known that *Robinson Crusoe* was largely based upon the true story of Alexander Selkirk, who had been abandoned on the uninhabited island of Juan Fernandes in 1704, and had lived alone there for more than four years. However, there were many other sources of Defoe's book, some true tales and some fictional, and he had adapted them for his purpose. Shakespeare's *The Tempest* was one of these sources, and although it would be an exaggeration to equate Prospero with Robinson Crusoe, and Caliban with Man Friday, the play did touch on the relations between civilized man and the indigenous savage.

In Defoe's tale the figure of Robinson Crusoe is represented as a merchant, corrupted by the vices of his time and his own sinful nature, being purged by his experience of loneliness and forced to repent and enter into a new relationship with his God. He is finally redeemed by Good Works (like Everyman) in converting the savage Friday to Christianity. This religious motive in the story is overlaid by Defoe's account of the practical stratagems employed to subdue nature on the island, and how a civilized man uses his wit to survive and create a tolerable order in a wilderness. The latter thread of the story undoubtedly has served to ensure the great popularity of the book, which still endures all over the world. The hundreds and perhaps thousands of books that have derived from *Robinson Crusoe* are known as Robinsonards and are part of survival literature which celebrates human ingenuity and love of adventure.[13] The religious motive of the book, which derives from Defoe's Dissenting Christianity, and the emphasis he puts on the power of loneliness to bring people to repentance, is omitted from most Robinsonards.

It is of interest to inquire into the nature of Crusoe's guilt that required purgation. He wrote:

> I have been in all my Circumstances a *Momento* to those who are touch'd with the General Plague of Mankind, whence, for aught I know, one half of their Miseries flow; I mean that of not being satisfy'd with the Station wherein God and Nature has placed them; for not to look back on my primitive Condition, and the excellent Advice of my Father, the Opposition to which, was, *as I may call it* my ORIGINAL SIN; my subsequent mistakes of the same kind had been the Means of my coming into this miserable Condition.[14]

Assuming that Crusoe was, in fact, a projection of Defoe's own self, which seems highly likely, we see here the characteristic Dissenter's tendency to insist on his own past sinfulness – until he was saved by faith and by good works – a characteristic that led Swift to condemn him as, 'so grave, sententious, dogmatical a rogue that there is no enduring him'.

The paradox is that in writing *Robinson Crusoe* Defoe produced a novel with the theme of a hero at odds with himself and society, and hence very lonely when forced to endure his own company, but who is eventually happily integrated into society when he has purged his sinfulness. Swift, on the other hand, gives us Gulliver, a fairly happy and sociable man who, as the

result of his final voyage to the land of the Houyhnhnms, becomes utterly alienated from human society and in his loneliness
seeks solace in the company of horses.

It is little wonder that during the Victorian era various ver-
sions of *Robinson Crusoe*, and derivatives from it, such as the
absurd *Swiss Family Robinson*, became popular among the middle
class in Protestant countries where the ethic was continual expansion and colonialism; whereas *Gulliver's Travels*, while
appreciated as a fantastic adventure story and published in garbled
forms for children, produced outrage in critics such as Thackeray
because of the final tale of the effect of the Houyhnhnms on
Gulliver.

Gulliver's eventual loneliness bears close examination. Aristotle
in his *Politics* stated:

> The man who is isolated – who is unable to share in the ben-
> efits of political association, or has no need to share because
> he is already self-sufficient – is no part of the polis, and must
> therefore be either a beast or a god.[15]

This observation was taken up by St Thomas Aquinas, partly
to explain the lonely life of John the Baptist and St Anthony
the Hermit, as quoted earlier in this chapter. In Gulliver's last
voyage he encountered the Yahoos who were certainly beasts
displaying all the worst vices of humanity, as contrasted with
the Houyhnhnms who were god-like in their perfection. Gulliver
realized that he could not hope to emulate the latter in their
virtue and nobility, but that he was not as bad as the Yahoos,
nor did he want to associate with them. When he eventually
came home he found that other human beings, his wife
and family included, were too akin to the Yahoos for him to
relish their company. He was not part of the polis and did not
wish to be.

Gulliver's Travels can certainly be classed as a Robinsonard, and
indeed it was partly a satire on Defoe's book. It was published
seven years after the latter and has several features which mock
Defoe's work. The third book, 'A Voyage to Laputa . . .' has some
quite unique features and one of them refers to the loneliness
of old age.

As I mentioned in Chapter 1, Gulliver visits the island of Luggnagg where he is told that they have people, known as Struldbruggs, who are immortal, and at first he is thrilled and excited by the idea of immortality and says:

Happy nation where every child hath at least a chance for being immortal! Happy people who enjoy so many living examples of ancient virtue and have masters willing to instruct them in the wisdom of all former ages! But happiest beyond all comparison are these excellent Struldbruggs who being born exempt from that universal calamity of human nature have their minds free and disengaged without the weight and depression of spirits caused by the continual apprehension of death.[16]

Gulliver expressed his surprise that he had not observed any of these immortals at the Court, for he imagined that the Prince would have had several of them about him as advisers and counsellors. When he had spoken in glowing terms of all the wonderful things he would do if he were granted immortality, the natives of the island disabused him of his illusions by showing him the Struldbruggs, some of them hundreds of years old, who exhibited all the disabilities and diseases of extreme old age, both physical and mental, including that of feeling utterly lonely and socially isolated because they could not properly understand the contemporary language – 'thus they lie under the disadvantage of living like foreigners in their own country'.

Swift was, of course, confusing the diseases of later life (which must have been very serious in the eighteenth century) with age itself, but nevertheless he makes the point that we should aim to make the best of living in the social era in which we happen to be born, and not to hanker after continued life in an altered society that will become foreign to us. This aspect of loneliness has some relevance to the Greek myth of Tithonus. Eos, goddess of the dawn, loved Tithonus, a mortal man, so she petitioned Zeus that he might live forever. The plea was granted, but she had not asked for his continued youthfulness, so while she retained her rosy beauty her lover withered as he aged and became utterly lonely. In Tennyson's poem based on the legend, he has Tithonus say:

> Let me go: take back thy gift:
> Why should a man desire in any way
> To vary from the kindly race of men,
> Or pass beyond the goal of ordinance
> Where all should pause, as is most meet for all.[17]

It took the genius of Swift to point out that not only did age bring physical changes to the individual as with Tithonus but, as society was constantly changing, men and women might be left behind and lonely if they lived too long.

In conclusion we may say that the basic difference between Defoe and Swift regarding their approach to loneliness was that the former regarded it as entirely reprehensible and remediable, whereas the latter did not. Defoe would have the solitary individual reintegrated into a decent God-fearing society such as he and his fellow Dissenters hoped to create; he recognized that there was much corruption in social institutions, and this meant that each individual should strive all the harder in a Bunyan-like manner to remedy it. Swift, on the other hand, saw social institutions as the joint expression of Yahoos acting in concert with a thin and often hypocritical veneer of civilization, and hence irredeemably corrupt. Only by standing aside from society could the individual preserve some decency and integrity in Swift's eyes. Both men were satirists and, to some extent, rebels against the existing social order, but their standpoints were basically opposed. While Defoe was the forerunner of nineteenth-century belief in progress through colonialism, steam-power and orthodox Christianity, Swift's influence led to such novelists as Samuel Butler and Oscar Wilde in the late nineteenth century and George Orwell in the twentieth. These latter writers believed that loneliness was an essential attribute of the human condition.

Charles Dickens

Of all the novelists of the nineteenth century, Dickens may be said to have appealed to, and represented, the greatest spectrum of society, and to have been the spokesman for his age. There are many other important nineteenth-century novelists, but none other than he have left their name in a meaningful adjective –

Dickensian – which is understood by everyone. His influence on other writers worldwide has been enormous: Fedor Dostoevsky was so influenced by him that he actually used Dickensian characters in some of his novels with their names little altered (Little Nell – Nellie Valkovsky; Steerforth – Stavrogin in *The Devils*).[18] Kafka, although with a style quite unlike Dickens's, was moved to write *The Trial*[19] after reading *Bleak House*. Leo Tolstoy too acknowledged Dickens's influence, and based his semi-autobiographical book *Childhood* on *David Copperfield*. Although many of Dickens's characters are simply caricatures, we sometimes encounter people in real life who strike us as being Dickensian.

Dickens was fascinated yet repelled by the subject of solitude and loneliness, and why this was so has never been satisfactorily explained. He wrote of himself as having been a lonely, bookish child, a spectator rather than a participant in children's play. We all know about his unhappy childhood experience of being withdrawn from school and sent to work in a blacking factory, but this can hardly account for his lifelong morbid preoccupation with loneliness. There is still some secret about Dickens's early experiences: in the short autobiographical piece he entrusted to his friend John Forster he described how he had been sent to work in the factory until his father, on his release from the debtors' prison, had insisted that he cease such menial employment and go to school. He wrote:

> I do not write resentfully or angrily for I know how these things have worked together to make me what I am, but I never afterwards forgot, I shall never forget, I never can forget, that my mother was warm for my being sent back.[20]

His mother's wish that he should go back to the blacking factory was apparently so that he should contribute seven shillings a week to the impoverished family budget. Dickens's brooding resentment (despite his denial of it) may have had something to do with the fact that there appeared to have been considerable conflict between his father and mother, which soured his childhood.

Bound up with Dickens's horror and fascination with loneliness was his preoccupation with prisons, partly, of course, because

his father had been in a debtors' prison. Prisons feature prominently in his novels and other writings. He visited prisons both
in Britain and America and was utterly appalled at the system of
solitary confinement which was then in force, and, with his own
half-belief in the supernatural, feared that such solitary incarceration would engender in prisoners the terror of ghostly
visitations:

> At Pittsburg I saw another solitary confinement prison. . . . A
> horrible thought occurred to me when I was recalling all I
> had seen, that night. What if ghosts be one of the terrors of
> the jails? I have pondered on it often since then. The utter
> solitude by day and night; the many hours of darkness; the
> silence of death; the mind for ever brooding on melancholy
> themes, and having no relief . . . imagine a prisoner covering
> up his head in the bedclothes and looking out from time to
> time with a ghastly dread of some inexplicable silent figure
> that always sits upon his bed, or stands in the same corner of
> his cell. The more I think of it the more I feel that not a few
> of these men are nightly visited by spectres.[21]

His attitude to aloneness is expressed in his piece entitled *Tom
Tiddler's Ground*. There he has his character Mr Traveller go to
an inn where the landlord tells him that there is a hermit in
the district, and asks him what he supposes a hermit to be, and
he replies:

> 'I'll tell you what I suppose it to be,' said the Traveller. 'An
> abominably dirty thing. . . . A slothful, unsavoury, nasty re-
> versal of the laws of human nature, and for the sake of God's
> working world and its wholesomeness, both moral and physi-
> cal, I would put this thing on the treadmill (if I had my way)
> wherever I found it either on a pillar or in a hole; whether
> on Tom Tiddler's ground or the Pope of Rome's ground, or a
> Hindoo fakeer's ground, or any other ground.[22]

There is no doubt that this speech represents Dickens's own
opinion about those who separate themselves from the world,
as is evident in many examples in his writings.

In Dickens's novels there are many lonely children: Florence Dombey, Little Nell, Oliver Twist, Jo the crossing-sweeper, Pip, Jenny Wren, David Copperfield. He writes with sympathy and considerable insight when he is writing of the loneliness of children, but when he writes of lonely old people he is sometimes remarkably brutal and often makes fun of the physical handicaps that come with age as though such disabilities served the victims right. He has a few characters who are neither children nor truly adults, who are lonely because they are the victims of circumstances – Amy Dorrit and Smike.

There are plenty of examples of those who are loners in old age, and Dickens generally makes them horribly ugly and vicious. He visits on them the stigmata that may come with ageing: some have lost the use of their legs (Grandfather Smallweed, Mrs Skewton, Silas Wegg), and some are afflicted with deafness (Peg Sliderskew, Grandmother Smallweed). Dickens mentions red-rimmed eyes, shrivelled skin, and a deformed body as being the accompliments of lonely old people who are in some way evil (Mrs Brown, Mrs Skewton, Fagin, Daniel Quilp, Ebenezer Scrooge, Arthur Gride). It is a fact that the unlovely and disabling effects of old age afflict both the just and the unjust alike, but Dickens reserves them for the unjust. The 'pleasant' old people (the Cheerable brothers, Mr Boffin, Mr Jarndyce) are spared the physical ravages of age, but they are far outnumbered by the unpleasant old people with their various ugly deformities.

Loneliness in later life appears to Dickens to betoken moral fault of some kind: avarice (Mrs Clennam, Scrooge, Fagin, Gride, etc.) or pride (Miss Haversham), but he does not attempt to show that loneliness may be the *result* of physical disabilities such as deafness and loss of mobility. He appears to write with glee when he describes his horrible old people. Mrs Brown, who attempts to pander for her daughter, is described thus:

> A very ugly old woman with red rims around her eyes, and a mouth that mumbled and chattered of itself when she was not speaking. . . . She had lost her breath, and this made her uglier still, as she stood trying to regain it: working her shrivelled yellow face and throat into all sorts of contortions.[23]

Fagin, who recruits street boys for theft, is introduced thus:

> A very old shrivelled Jew, whose villainous-looking and repulsive face was obscured by a quantity of matted red hair. He was dressed in a greasy flannel gown with his throat bare.[24]

Peg Sliderskew appears as: 'A short, thin, weasen, blear-eyed old woman, palsy-stricken and hideously ugly'.[25] She is also profoundly deaf. One might quote endlessly the grisly descriptions that Dickens gives of his lonely old people which mingle graphic details of physical deformity with moral defect. Where there is no obvious moral deficiency but the characters are merely old, lonely and unfortunate, as in case of Mrs Gummidge, the 'lonelorn creetur', and those who are demented – Grandmother Smallweed and Mr F's Aunt – Dickens holds them up to ridicule.

Towards the end of the present century, with the emergence of the Third Age, many writers have protested at the negative and inaccurate public image of older people that we have inherited from the past, such as has been described in Chapter 2.[26] The writings of Dickens with their popularity and unforgettable characterization have undoubtedly been one of the factors that have helped to maintain this negative stereotype. Such individuals as he described unquestionably did exist and will continue to exist, but in appreciating this fact we must remember that they are as far from the norm as his very occasional and rather unbelievable old characters like Mr Boffin who are models of generosity and cheerfulness.

Joseph Conrad and Graham Greene

These two novelists represent an outstanding contribution to the literature of loneliness in the twentieth century. In Conrad's writings, both the short stories and the novels, he centres on the problem of loneliness perhaps more than any other later writer. Greene is more diverse in his various works, but when he writes of situations in which the loneliness of individuals is prominent he is very much influenced by the earlier writer who was his master.

Conrad's view of loneliness is a very grim one: not for him

the possible dignity of solitude; in his view solitude is equiva-lent to an unpleasant loneliness, almost invariably the product of some moral weakness and guilt on the part of the solitary person. Adam Gillon[27] refers to most of Conrad's principal charac-ters as 'isolatoes', a term used by Herman Melville in his *Moby Dick* to characterize men who 'not acknowledging the common continent of men, but each isolato living on a separate conti-nent of his own'. According to the critic Ben Mijuskovic,

> Conrad . . . is one of the most subjective of English writers, and he is so precisely because he is convinced that each man is alone and a stranger even to himself. Blackness, darkness, blindness, even the jungle, 'where men deteriorate in soli-tude' are mere symbols for the individual's awareness of his own absolute isolation from his fellows and from his surround-ings, the jungle or the sea, the latter two elements representing an untimate realm of uncaring being set in opposition to in-volved and involuted consciousness.[28]

Conrad's preoccupation with loneliness following a particular pattern of betrayal is chiefly expressed in his early work – *An Outcast of the Islands* (1896), *The Nigger of the 'Narcissus'* (1897), *Heart of Darkness* (1898), *Lord Jim* (1900). In his first decade of writing he is strongly dominated by his pent-up experience of frustration and his sense of guilt at having deserted his mother-land. By 1902 or thereabouts he imagined that he had become 'a burnt-out case', just as Graham Greene was to regard himself in 1960. The term 'a burnt-out case' is used by Greene initially to indicate a sufferer from leprosy where the disease is in re-mission having left the patient more or less maimed.

Conrad, in his 'Author's Note' to *Nostromo*, wrote:

> I don't mean to say that I became conscious of any impend-ing change in my mentality and in my attitude towards the tasks of my writing life. And perhaps there was never any change, except in that mysterious extraneous thing which has nothing to do with theories of art; a subtle change in the nature of inspiration; a phenomenon for which I cannot in any way be held responsible. What, however, did cause me

some concern was that after finishing the *Typhoon* volume it seemed somehow that there was nothing more in the world to write about.[29]

It was as though having fully stated his case – his terribly pessimistic conclusion that man's idealism was futile and that he must remain a lonely soul, at best performing his duty in a pre-ordained framework of discipline – he had no more reason to write.

Some of the better-known writers other than Greene have been influenced by Conrad when they have dealt with the theme of loneliness, even when in general their writings have not resembled his. Evelyn Waugh has taken the lonely Mr Kurtz from the African jungle and put him in the jungle of South America (the scene of Conrad's *Nostromo*) in the person of Mr Todd in *A Handful of Dust*.[30] There the resurrected Mr Kurtz, instead of being obsessed with 'unspeakable rites' and a passion for ivory, solaces his lonely old age by being read the novels of Charles Dickens by the 'Marlow' figure – Tony Last – whom he has literally captured. Mr Todd is the son of a Conradian stock figure – an expatriate European man who has gone to seed in the wilderness. Waugh has indeed followed Conrad's analogy completely in comparing the savagery of the wilderness with the savagery that he had experienced elsewhere in the civilized world. Roland Reinert comments:

> London and the jungle are but two facets of the same universe: in both Tony [the Marlow figure] appears as a victim who is gradually deserted by those around him, and there is a moral point to this symbolic identification of the two worlds. Civilization and savagery appear as interchangeable concepts; the so-called civilized members of society are no better than the South American savages: to the physical savagery of the latter corresponds the moral savagery of the former.[31]

In Chapter 1, I referred to loneliness as either stemming from the circumstances surrounding an individual or being an inborn trait in that person; in Conrad's case both factors appear to have dominated his life.[32] The son of a Polish aristocrat who was exiled from his native land to Russia, he grew up in a rather mournful household bewailing its isolation. Orphaned at the age of 11 he came under the guardianship of an uncle, and perhaps in an attempt to escape from his rather miserable background, he elected to leave home at the age of 17 and to seek adventure as a sailor. He did indeed separate himself from his Polish background, but in doing so he encumbered himself with a guilty feeling of having deserted his homeland, when as the son of a noble Polish family, it was his duty to continue to strive for its freedom from the Russian yoke and uphold its honour. This theme of betrayal is reflected in many of his writings. More than thirty years after he had left the sea he became a well-known English writer, but he remained an 'isolato', writing in excellent English yet continuing to speak in a thick Polish accent which surprised many of those he met. Perhaps this habit of speech was retained to emphasize the fact that he was not really an Englishman but an everlasting foreigner.

Conrad was a shy man who actually lied about himself to mislead those who were interested in his biographical details, but his personal problem of loneliness comes into many of his stories which are highly autobiographical. An excellent picture of him is conveyed by Bertrand Russell in his *Autobiography*:

> He was, as anyone may see from his books, a very rigid moralist and by no means politically sympathetic with revolutionaries. He and I were in most of our opinions by no means in agreement, but in something very fundamental we were extraordinarily at one.
>
> My relation to Joseph Conrad was unlike any other that I have ever had. I saw him seldom, and not over a long period of years. In the out-works of our lives we were almost strangers, but we shared a certain outlook on human life and human destiny, which, from the very first, made a bond of extreme strength. . . . Of all he had written I admired most the terrible story called *The Heart of Darkness*, in which a rather weak idealist is driven mad by horror of the tropical forest and loneliness among savages.[33]

The surprising friendship between Russell the revolutionary thinker and Conrad the arch-Conservative may have owed

something superficially to the fact that they were both aristo-
crats who retained an aristocratic view of society for the whole
of their lives, but at a deeper level both had an aching sense of
loneliness although they had plenty of social contacts, and in
this lonely quest for peace of mind they had a common problem. In the end Russell was able to come to terms with his quest,
as he describes in the long passage that has been quoted in Chapter
1, but Conrad was not so fortunate, and in his story *Amy Foster*[34] he depicts as tragic and lonely an end to life as that of
Kurtz in *The Heart of Darkness*.[35]

In *Amy Foster*, Yanko, a peasant from Eastern Europe, is ship-
wrecked on his way to America, and lands up in a Kentish village
where everyone resents his 'foreign-ness' and makes him feel an
outcast – much as Conrad himself may have felt when he first
left Poland. However, Amy, a plain and simple English girl, takes
pity on him and befriends him, eventually marrying him and
having a child by him. In the end he succumbs to a fever, and
in his delirium he reverts to his native language which is totally
incomprehensible to his wife, and Amy becomes terrified by his
strangeness and, snatching up her child, she leaves him to die
as lonely a death as Jude, Thomas Hardy's character in *Jude the
Obscure*. It is noteworthy that Conrad also married an English
woman and had children by her, but with his extreme pessimism he may have feared such a lonely end to his life.

The archetype of Conrad's lonely man is Lord Jim who is the
'hero' of the novel of that name. Jim does indeed set out to be
a noble hero, to achieve much and to be admired by his grateful
fellow-men (women play little active part in Conrad's books),
but due to his own cowardice and moral weakness he finds that
eventually all he achieves is loneliness:

> a hard and absolute condition of existence; the envelope of
> flesh and blood on which our eyes are fixed melts before the
> out-stretched hand, and there remains only the capricious,
> unconsolable, and elusive spirit that no eye can follow, no
> hand can grasp.[36]

But what if a man does cling to his ideals and by self-discipline
achieves what he conceives to be his true role in life rigidly

adhered to, like Captain McWhirr in *Typhoon*? Such a figure has
to embrace a different sort of loneliness, a loneliness that may
affect many in later life: the loneliness of 'the tyranny of command'. During the tempest the crew obtain some comfort from
the presence of their resolute captain on deck, but McWhirr 'could
expect no relief of that sort from any one on earth. *Such is the
loneliness of command.*'[37]

We are reminded of the return of Ibsen's Peer Gynt, rich and
apparently successful, but envious of the poorest of sailors who
had humble homes to go to. Peer had not obtained worldly success by clinging to ideals like Captain McWhirr, but by his
cleverness and ruthlessness, yet such was his loneliness in old
age that he doubted the validity of his own existence and feared
to be melted down by the figure of the Button Moulder: he even
petitioned Satan to find him a home in Hell. But Ibsen was a
stronger character than Conrad, and at the end of his play *An
Enemy of the People*, he has Dr Stockman declare with evident
satisfaction, 'The strongest man on earth is he who stands alone.'

Greene does not bring the theme of loneliness into his books
as insistently as Conrad, but when he does they have a distinctly
Conradian atmosphere. He was quite aware of the influence of
his fore-runner, and indeed he struggled against it. In 1961 he
wrote:

> Reading Conrad – the volume called *Youth* for the sake of
> *Heart of Darkness* – the first time since I abandoned him about
> 1932 – because his influence was too great and too disastrous.
> The heavy hypnotic style falls around me again, and I am
> aware of the poverty of my own. Perhaps now I have lived
> long enough with my poverty to be safe from corruption.[38]

But Greene was never free of Conrad's influence when the theme
of loneliness was concerned. For his two novels which deal with
loneliness most prominently, *The Heart of the Matter*[39] and *A Burnt-
Out Case*,[40] he turns to Africa and his memory of how, as a young
man, he imitated Conrad by travelling up the Congo, and like
his master, contracted a near-fatal illness. He was to return to
Africa three more times, the last to obtain material for *A Burnt-
Out Case*. The influence of *Heart of Darkness* on these two novels

is too obvious to need pointing out, although Greene puts a
religious gloss on them, as on his many other books. For Conrad
loneliness was not necessarily a matter of religion; his approach
was more existential.

Above, the theme of loneliness has been attributed mostly to
the first decade of Conrad's writing, but it comes in very powerfully in one of his later works, *Victory*,[41] which was published in
1915. Here the main character, Heyst, is a Swedish nobleman;
perhaps this is an oblique autobiographical reference to the lonely
author, a Polish nobleman. Here Conrad has apparently abandoned his own feeling of guilt, and he makes Axel Heyst just
lonely by nature and seeking solitude for its own sake. There is
plenty of evil in the book and it is mainly projected on the
character calling himself Mr Jones, a cultured English gentleman
who has been hounded out of England for some unspeakable
crime, to lead a lonely and criminal career wandering the fringes
of the civilized world. Mr Jones is a pathological woman-hater
who murders his criminal companion, Ricardo, in a jealous rage
for deserting him for a woman. There is a strong hint here that
the gentlemanly Mr Jones is a lonely outcast because he is a
homosexual, and in the England of 1915 homosexuality was
regarded with such abomination by the public that this is a
possibility. Indeed, did Conrad mean to imply in his novels that
his lonely wandering figures were similarly inclined? Kurtz had
an 'intended', a cultured lady who lived in London, but he appears to have continued his solitary life in Africa practising
'unspeakable rites' rather than returning to marry her.

The question of Conrad implying homosexual leanings to a
number of his 'isolatoes' has been discussed by various critics.
In comparing his novel *Chance* to *Victory*, Adam Gillon comments:

> Another basic similarity between the two novels is Conrad's
> interest in sensuality, more specifically of the homosexual
> variety.... The effeminate pistol-packing Jones is described
> by his 'secretary' Ricardo as a *freakish* gentleman who hates
> women. 'Yes the governor funks facing women'. Eventually
> the 'governor' will shoot Ricardo for his involvement with a
> woman.[42]

Indeed, it has been suggested by Jeffrey Myers that Heyst is homosexually inclined also,[43] and other critics have commented on the homosexual element in others of Conrad's novels.[44] The theme of homosexuality leading to loneliness was quite natural in the earlier part of this century because it was so universally condemned in this country, and when it found expression in literature, as in Ratclyffe Hall's *Well of Loneliness*,[45] it could lead to a criminal prosecution. It is of interest that even though public attitudes to homoerotic love have now been totally liberalized the same theme of it leading to loneliness still finds expression in literature, as in Tom Wakefield's novel *Mates*.[46]

Greene's dependence on Conrad's *Heart of Darkness* for both *The Heart of the Matter* and *A Burnt-Out Case* is unquestionable. It hardly needs pointing out that Greene's later novel is simply a re-hash of his former book, with the principal characters re-named and new situations introduced. What is of greater interest is that not only does *Heart of Darkness* set the scene, but Conrad's later novel *Victory* provides much of the plot for *A Burnt-Out Case*. Both Heyst and Querry are lonely creatures seeking solitude, but their sense of pity remains and leads them to become involved with young women who fall in love with them, and such involvement leads to their death. It is perhaps easier to believe in Heyst; whether Conrad intended him to be a repressed homosexual or not does not matter: he shows some development of character. First, his pity is shown towards another man, Morrison, but this ends in tragedy, Morrison's death, and is the cause of unjustified rumours being spread about Heyst, and he flees from human companionship. He then shows pity to a young woman and rescues her from an intolerable situation; she falls in love with him and tries to protect him but is the cause of his death. At last his death is due to a true, human action. Why Greene's Querry seeks solitude is less convincing: he has indeed a cause for guilt – his brutal womanizing in the past – and thus he is like the isolatoes of Conrad's earlier books. He too shows pity for a young woman in an intolerable situation; he rescues her, she falls in love with him and he meets his death at the hands of her jealous husband.

George Orwell

George Orwell is now known throughout the world because of the success of his last two books, *Animal Farm*[47] and *Nineteen Eighty-four*.[48] His name, like Dickens's, has given rise to an adjective, Orwellian, which implies the sort of satire which he directed against totalitarian institutions. The words and phrases which he used have now become incorporated into common speech: 'Newspeak', 'double-think', Big Brother', 'some being more equal than others', etc. He has been been compared with Jonathan Swift in his satiric approach to criticisms of society and political institutions, and indeed there are many similarities between the two men. In their writing they are memorable for creating the lonely figures of Gulliver (after his visit to the land of the Houyhnhnms) and Winston Smith, the latter being a lonely misfit in the monstrous society of his time.

We have an interesting contrast between the satires of Swift and Orwell: Gulliver started off as a normally sociable man but at the end of the fourth book he became completely solitary by choice, utterly cut off from his fellow humans because he was revolted by what he perceived as their moral and physical degradation, so reminiscent of the Yahoos. Winston Smith, on the other hand, began as a lonely individual who was not integrated into the society of Oceania, but ended as someone who was only too well integrated, after the Thought Police had done with him, so that he loved Big Brother. The reader is left to ponder whether it is preferable to be a lonely Gulliver, unhappy in his solitude but dignified and his own man, or Winston Smith, well integrated into the society about him and caring deeply for its professed ideals, but utterly a cipher without a will of his own.

Reviewing Orwell's *Nineteen Eighty-Four* Frederic Warburg wrote:

> This is amongst the most terrifying books I have read. The savagery of Swift has passed to a successor who looks upon life and finds it becoming ever more intolerable. . . . Orwell has no hope, or at least he allows his readers no tiny flickering candle of hope. Here is a study in pessimism unrelieved, except by the thought that if a man can conceive *Nineteen Eighty-four* he can will to avoid it.[49]

Various critics said much the same about Orwell's book and he replied as follows:

> I do not believe that the kind of society I describe necessarily *will* arrive, but I believe (allowing of course for the fact that the book is a satire) that something resembling it could arrive. . . . The scene of the book is laid in Britain in order to emphasise that the English-speaking races are not innately better than anyone else and that totalitarianism, *if not fought against*, could triumph anywhere.[50]

Another interesting contrast between Swift and Orwell concerns the role of the Yahoos and proles in their books. Unlike many Leftists of his time Orwell made no attempt to glorify the working class; he depicted them as he saw them, and it must be remembered that in the England in which he grew up there was some truth in the saying which he quoted as having been said to him in childhood (but did not himself endorse), 'The lower orders smell.' He depicted the proles in all their dirt, shabbiness and disorder, living in squalor and warm matey-ness; they were in fact Yahoos. Yet, speaking through Winston Smith, he made the prediction 'If there was hope it lay in the proles! . . . The future belonged to the proles.' A number of critics have debated whether or not Orwell himself meant this; certainly his bitterest satire was directed towards his own class of people – the middle-class intelligentsia. But if Swift's Yahoos are proles, what of the clean, well-ordered and highly moralistic Houyhnhnms – are they the ruling Party intelligentsia? In his review of *Gulliver's Travels*[51] Orwell referred to the Houyhnhnms' society as being 'totalitarian', and as far as I know no critics have debated this, but of course the question arises of whether it is better to be an aloof intellectual, risking the emotional loneliness of seeing one's society all too clearly, or be an unthinking prole, muddling along in the mateyness of mass culture.

The question of loneliness in later life was recognized by Swift, as discussed earlier, in his concept of the Struldbruggs who lived on and on, becoming ever more lonely as the society about them changed. Orwell hardly mentions older people at all in his books and essays, but in *Nineteen Eighty-Four* he has Winston go into a pub in a proletarian district in search of clues as to what it was

really like in the past before the revolution. He tries to interro-
gate an old man but gets little satisfaction because the fellow is
rather silly and incoherent. This old man gives signs of experi-
encing the same sort of isolation as the Struldbruggs – society
has changed a great deal and he has not. His complaints are
about the most mundane things such as the fact that beer is no
longer sold by the pint, which suited him, but by the litre.

Orwell's previous book, *Animal Farm*, was equally as famous,
partly because it was largely a satire on Stalinist Russia, and as
the Cold War developed it was used as anti-Soviet propaganda.
Like Swift's fourth Gulliver book, it was a Beast Tale, but instead
of making the animals – horses – superior to mankind Orwell
chose pigs in the role of oppressors over all other animal species
on the farm. In the end the pigs are shown to be as corrupt and
ignoble as the humans who had exploited the farm animals – in
fact they represented the Bolsheviks in the Soviet Union. The
same ideas are expressed in Orwell's final two books: that power
corrupts and that the common people are dominated not just
by physical violence but by an elaborate system of lies and humbug.
Again Orwell introduces a lonely figure who stands outside society
and is not fooled by propaganda and a need to go with the herd
– Benjamin the donkey. Benjamin is not an important figure in
the book, but some critics have identified him as representing
the author himself. One commentator, however, Patrick O'Brian,
strongly opposes this interpretation, pointing out that Orwell was
not a cynic but a very active and generally constructive critic.[52]

Surprisingly Orwell wrote very favourably of Charles Dickens's
work and claimed that he had been a formative influence in his
development as a writer. Unlike Orwell, Dickens was not in the
least a political writer and was somewhat smugly middle-class in
his ideals. The attraction that he had for Orwell was because he
was something of a radical and satirist, but chiefly that he stood
for the ordinary decency that Orwell perceived as of paramount
importance in the Common Man. Orwell also felt a kinship with
Dickens because of the latter's repeated depiction of the loneliness of children and their abuse at the hands of adults,
remembering his own lonely experiences as depicted in *Such,
Such Were the Joys*[53] but he had nothing to say about the loneliness of older adults.

Loneliness in post-Orwellian literature

In his book *George Orwell: a Literary Life*, Peter Davison writes the following:

> great writers spawn industries dedicated to the examination of their work and the furtherance of their academic careers. There is a Shakespeare industry, a Joyce industry and, as a mark of his stature I suppose, an Orwell industry. . . . There is even a *Nineteen Eighty-Four* industry. Much that has been written on this novel, its origins and implications is excellent and genuinely helpful.[54]

It is also true to say that there has been, and still is, an anti-Orwell industry dedicated to the disparagement of his work and to the expungement of his influence on present and future literature. The earliest and most vociferous opposition came from the Communists and the 'fellow-travelling' literati – not members of the Party themselves, but devoted to hushing up the worse atrocities of the Communist regime and serving as apologists for Marxist-Leninist ideology. This element of the anti-Orwell industry has largely passed away with the waning of the Communist Party internationally. A second strand of opposition has come from the growing radical feminist movement which began to be prominent in the 1950s and has increased its influence on literature as time has gone on.

Feminist literature tends to represent loneliness as the result of the intolerance of male-dominated society on women who wish to be free to choose their own lifestyle whether it is heterosexual, homosexual or entirely a-sexual, rather than accepting the role that men are supposed to demand of them. Many feminists will not easily forgive Orwell for making Julia, in *Nineteen Eighty-Four*, a rebel 'only from the waist downwards'. A continuing opposition to the effect of Orwell's writing on modern literature comes from an 'upper-crust' of academic literati who are not political but deem it meet to regard the canons of good writing far removed from the judgement of the common reader. In the book already mentioned Davison relates how a meeting of American professors of English in 1983 decided that Orwell was a 'journalist' and a 'didactic writer' who 'failed to live up to top literary standards, with *Nineteen Eighty-four* in particular lacking in literary sophistication'. But Orwell wrote for the common reader, in prose that is admirable in its clarity, as did Swift and Dickens, and his books and essays will continue to be read long after these professors and their works are forgotten.

It cannot be claimed that Orwell has had much *positive* influence on writers after his time, particularly in relation to the literature of loneliness. His influence has been *negative* however, in that few people have attempted to write utopias because of the lasting shadow of *Nineteen Eighty-Four*. Aldous Huxley and B.F. Skinner published utopias in the post-Orwellian period, but both novels were failures and soon forgotten.[55] The lonely figure of Winston Smith remains in the public mind as a reminder of what might happen to each and all of us if we do not successfully oppose the forces that make for the totalitarian state.

Postscript

At the beginning of this chapter I pointed out that it would refer to the general topic of loneliness in literature and not specifically to the loneliness of older people. This has been necessary because, as pointed out in Chapter 2, very little has been written about the lives of older people at all. Their lives have not been deemed sufficiently interesting or important to warrant novelists and others making them the subject of their work. Now with the coming of the phenomenon of the Third Age, matters are changing and a new genre of literature is emerging as I have discussed elsewhere.[56] As far as the loneliness of the elderly is concerned, I must repeat what I have already pointed out – it is important to realize that the things that make people lonely are really the same in childhood, mid-adulthood and old age. We are the same people in our later life, although we have to cope with new problems that come with new situations developing with age.

5
The Benefits of Solitude

The Chinese sage Ching Chow proclaimed that 'What fools call loneliness, wise men know as solitude'.[1] We do not know, of course, precisely what he meant by 'fools': did he mean people of every level of culture and education who are foolish in their outlook and behaviour, or had he in mind those who are simple and unlettered like the average Chinese peasant of his time? Again, what exactly did he mean by 'wise men' – those who are educated and learned, or those who are sensible and canny whatever their status in life? It is certain that the educated have the advantage of having abundant intellectual resources within themselves which they can draw upon when quite cut off from human companionship, but there are plenty of examples of men and women of little education who have happily embraced a solitary way of life in occupations such as shepherding or religious seclusion. Perhaps the individual's reaction to being alone is more a matter of personal temperament and the factors that have determined the aloneness.

Well-known people throughout the ages have echoed the Chinese sage's dictum quoted above: 'A wise man is never less alone than when he is alone' – (Jonathan Swift);[2] 'Solitude is the best nurse of wisdom' (Laurence Sterne);[3] 'Solitude is the nurse of enthusiasm, and enthusiasm is the true parent of genius. In all ages solitude has been called for – has been flown to' (Isaac D'Israeli);[4] Poets too have sung the praises of solitude, as in James Thomson's 'Hymn on Solitude':

> Hail mildly pleasing Solitude,
> Companion of the wise and good;
> But from whose holy piercing eye
> The herd of fools and villains fly.
> Oh! how I love with thee to walk,
> And listen to thy whispered talk,
> Which innocence and truth imparts,
> And melts the most obdurate hearts.[5]

The naive and much parodied poetry of James Thomson which celebrated the cult of solitude in the eighteenth century nevertheless influenced later major poets, including Wordsworth, who often echoed the theme, as in his poem *A Poet's Epitaph*:

> The outward shows of sky and earth,
> Of hill and valley he hath viewed;
> And impulses of deeper birth
> Have come to him in solitude.[6]

Imposed solitude

Those who wrote in praise of solitude were, of course, writing of that which is voluntarily embraced. Where solitude is enforced as in imprisonment or banishment, it generally results in acute loneliness. The solitary confinement of prisoners is recognized as being a very harsh punishment, and the desperate loneliness which it engenders tends to break down the victim's normal personality. It has been used by repressive regimes in order to secure confessions, and to secure signed confessions. For some people the loneliness of being isolated in prison makes any human contact rewarding, and so those who are detained for long periods may welcome even the company and conversation of their interrogators. It is understandable that they may greatly value the periods when they are taken out for questioning, even forming a relationship with the interrogator. The latter may find it useful to pretend to be sympathetic with the prisoner's plight, and according to Hinkle and Wolff, who investigated the practices used in Stalinist Russia:

There are instances of prisoners who signed depositions largely out of sympathy with their interrogators, because they felt that these men would be punished if a proper deposition were not forthcoming. In other words, the warm and friendly feelings which develop between the prisoner and interrogator may have a powerful influence on the prisoner's behaviour.[7]

The so-called 'brainwashing' techniques that the Chinese employed with their American prisoners during and after the Korean War depended largely on the effects of imposed solitude interrupted with occasional sessions with their Communist captors who managed, in some cases, actually to change their prisoners' political and social orientation. Arthur Koestler, who was a prisoner of the Franco forces during the Spanish Civil War, used his personal experience of incarceration in writing his novel *Darkness at Noon*, which deals with the breakdown of an Old Bolshevik and his acceptance of the new policies and ways of thinking of the Stalinist state during the great purges in Russia.[8]

Koestler's experience of imprisonment was not very long and he was not subjected to any 'brainwashing' techniques, but it led him to study the effects of solitude and to write of it producing

a feeling of inner freedom, of being alone and being confronted with ultimate realities instead of with your bank statement. Your bank statement and other trivialities are again a kind of confinement. Not in space but in spiritual space. . . . So you have got a dialogue with existence. A dialogue with life, a dialogue with death.[9]

People vary a good deal as to how they react to such enforced solitude and coercive methods, and there are some accounts of individuals who have been remarkably resistant to them. Such a one is Dr Edith Bone who was imprisoned for seven years in Communist Hungary.[10] She was a highly gifted linguist, an active member of the Communist Party who went to Hungary to translate English scientific books. She came under the false suspicion of being a British spy and was arrested in 1949. She was at that time over 60 years of age. Vehemently protesting her innocence, she resisted all attempts to make her sign a false confession.

Dr Bone spent seven years in prison, for much of the time in complete solitude, denied access to books or writing materials, and under dreadful physical conditions. She maintained her sanity partly by constantly treating her captors with contempt and never showing any signs of wanting their company, withstanding the harshest of privations rather than wavering in her attitude of complete defiance.

Edith Bone's book is a remarkable account of the nature of solitude, showing how someone with a well-stocked mind may overcome the deleterious effects of such a situation in which there is an almost total absence of stimulation from without. With remarkable self-discipline she kept herself active, even when deprived of books and writing materials, with translating poems she remembered into foreign languages, and similar mental exercises. But, most important, she put solitude to good use in rethinking her attitude to Marxist-Leninism, the cause that she had embraced for many years. She had been an active member of the Communist Party at the time when Russia had been under the domination of the Stalinist regime, and an influential apologist in the West for all the worst abuses that went on there. In the period of her solitude in prison she was able to come to terms gradually with the fact that she had been a self-deceiver and had used her considerable talents continually to mislead others about the realities of the situation in the USSR. Nagging doubts about the palpable lies in Soviet propaganda had always worried her, but in her very busy life she had been able to brush them aside; it took the long-continued solitude of prison really to examine herself and to effect a fundamental change in her outlook which led to her renouncing her Communist allegiance.

A similar, although less remarkable case, is that of Christopher Burney who, during the Second World War, underwent a period of 18 months of solitary confinement in France in a prison controlled by the Germans during the Occupation, as described in his book.[11] When he had adapted to solitude, he experienced a state of calm peace of mind so that towards the end of his incarceration when he had the opportunity of communicating with other prisoners he felt no real need to, and even resented, to some degree, their well-meant efforts to get in touch with him. He was eventually transferred from the prison to the concentration camp at Buchenwald. Here the great detachment that he had acquired in solitude enabled him to endure the horrors of camp life with remarkable resilience until he was eventually released by the advance of the American army.

Solitude in later life

It may seem strange to compare the conditions of solitude that tend to come in the later part of life with the dreadful situation of prisoners suffering from enforced solitude, but there are some interesting parallels. The solitude of the older person is, in a sense, 'enforced' in that it comes about through a number of conditions that are not voluntarily chosen but imposed through the process of ageing. Friends and family begin to die, and as was shown in the survey described in Chapter 3, the majority of women in their seventies are widows or separated from their husbands. It is not easy to make new friends in later life, partly because most younger people have not a great deal in common with the older generation, and there is a limited amount of socialization between the generations, except for family members. With the small families that have come into being this century owing to deliberate family limitation, older people have had fewer and fewer relatives. Another significant factor that produces solitude in later life relates to the disabilities that come with age: reduced mobility, deafness, less energy and less willingness to go out at night when younger people are socializing.

The influence of temperament

It was noted above that one important personal attribute that enabled people to tolerate extreme solitude was their temperament. Temperament is difficult to define and indeed to subject to any sort of measurement. The psychiatrist Carl Jung described a measure that is certainly relevant to people's adjustment to being alone: introversion–extraversion. The more extraverted person has a greater need for social stimulation and is dependent on the company of others to feel comfortable and well-attuned to life, whereas more introverted people are less dependent on such stimulation from without. It follows that more introverted people easily endure or actively enjoy solitude, whether imposed or voluntarily embraced.

Jung's use of the concept of introversion–extraversion was not entirely original. The term 'extraversion' first appeared in Dr Johnson's *Dictionary of the English Language* which was published in 1755, but Dr Johnson did not give the term the meaning it acquired later. J.A.H. Murray in the *Oxford Dictionary* of 1888[12] quotes the eighteenth-century lexicographer Coles defining extraversion as 'a turning of one's thoughts upon outward objects', and W.D. Whitney in the *Century Dictionary* of 1899 defined introversion as 'the act of introverting, or the state of being introverted; turning or directing inward, physical or mental'.[13] It will be apparent that the concept of introversion–extraversion has been around a very long time, and it is indeed possible that this is what the Chinese sage Ching Chow was getting at when he used the words that have been translated into English as 'fools' and 'wise men', although it would be quite mistaken to equate extraversion with folly and introversion with wisdom.

Although Jung's use of the concept has its place in Jungian psychiatry, the terms introversion and extraversion were seldom used in the psychology of the early twentieth century until the psychologist Hans Eysenck began his career of trying to make psychology a scientific subject in his researches of the 1940s. He refers to his early work thus:

> In 1947 I published my first book, *Dimensions of Personality*, and in doing so took under my wing a most unattractive old thing with a caricature of a face – to wit, the concept of introversion–extraversion. . . . To many, if not most, psychologists interested in personality it seemed as if I had attempted to resurrect a corpse – equivalent, perhaps to trying to reintroduce into physics the notions of phlogiston, or aether, or a geocentric planetary system.[14]

Eysenck's efforts were to prove highly successful, and he and his colleagues did a great deal of useful work in the clinical and experimental fields of psychology that was to become influential worldwide. One of his great contributions was to demonstrate that Jung's concept was not only valid and important but as a dimension of personality it was *measurable*. It is not the case that there are two exclusive types of people, 'introverts' and 'extraverts', but that it is a matter of degree how far each individual has a tendency to lean towards one or the other of the two extreme poles, and this must be borne in mind even though for the sake of brevity psychologists sometimes write of 'introverts' and 'extraverts'.

One of the ways of measuring the degree of individuals' tendency to act in one way or the other is to ask them to complete self-rating questionnaires. More extraverted people tend to say 'Yes' to many of such questions as – 'Can you usually let yourself go and enjoy yourself a lot at a lively party?', 'Do you often long for excitement?', 'Do you often do things on the spur of the moment?', 'Would you be very unhappy if you could not see lots of people most of the time?' More introverted people would say 'No' to most of these items, but would endorse many such items as, 'Do you stop and think things over before doing anything?', 'Do you prefer to have few but special friends?', 'Do you like the kind of work that you need to pay close attention to?' These are a few items taken from the *Eysenck Personality Questionnaire* which asks a great variety of questions of this nature.[15]

Superficially it might appear that little important information could be obtained by asking people to complete such questionnaires, but over the last half-century an enormous amount of important and productive research has been carried out using this and similar questionnaires, not only in clinical work but in a variety of other fields as well. One such interesting piece of research was determining the effect of measured quantities of alcohol on driving skill, in relation to the individual driver's position on the introversion–extraversion continuum.[16] More extraverted drivers are worse affected by alcohol in their skill, and if some people are very introverted, their driving may be slightly improved by a *small* dose of alcohol.

The relevance of this aspect of personality for our present discussion is that numerous research studies have shown that as a natural result of ageing most people tend to become more introverted in later life. Although many people would attribute this effect to social factors, Eysenck offers evidence that it is partly due to physiological changes in the nervous system.[17] The increase in introversion in later life is very fortunate for, as already noted, social factors generally ensure that we lead more solitary lives as we age, and a greater degree of introversion ensures that solitude is more easily accepted, and even enjoyed by more introverted people. There is also evidence which is discussed by Stuart-Hamilton[18] that age-related changes are different for men and women. In their late teens men are significantly more extraverted than women, but thereafter they show a more rapid decline in their level of extraversion, so that in the late sixties it is women who are more extraverted.

Older men living alone

As demonstrated in Chapter 3, there are relatively few men in later life who are unmarried and living on their own. When older men are bereaved they tend to take the loss of their partner harder than do women, and in the first year after bereavement there is a significant rise in the death-rate of widowers. When widowers have recovered from the sadness and shock of bereavement, most of them adjust very well to the solitude of living alone.

In the study of members of the University of the Third Age (U3A) reported in Chapter 3, only nine men were living alone and they reported a higher degree of loneliness than the men living with their wives. This may be compared with a study by Rubenstein who studied 42 men over 65 who were living alone in order to determine, among other things, how lonely they were if at all.[19] He found that 60 per cent of them reported that they were lonely to some degree, but the remaining 40 per cent said that they were 'Rarely or never' lonely. Reading Rubenstein's book one is impressed by the fact that many men had settled down perfectly contentedly to solitary living after bereavement, even though they did not have a great deal of contact with family or friends. In accepting the solitude of their condition the relatively high degree of introversion of older men is undoubtedly a favourable asset.

Older women living alone

As was shown in Chapter 3, many more women than men in later life live alone because of bereavement, and a significant minority have separated from their husbands or have never married. It is of interest to see the result of the relatively higher degree of women's extraversion in later life. They adapt to their solitary condition by becoming more gregarious and join bodies such as the U3A which present, as one of their main objects, opportunities for social mixing. By the age of 75 there are twice as many women as men in the population, but the ratio of women to men in the U3A is about 3:1. Women make friends more easily than men in the later years of life and although they may live in separate houses, they temper their solitude with a greater degree of social intercourse according to their individual preference.

Solitude versus the family nest

One of the persistent misapprehensions about the older generation is that they want to be more dependent, and closely in touch with, their younger relatives, and that if they are not they will be pathetically lonely. Many younger people simply cannot understand that their parents and other older relatives are perfectly happy living in relative solitude and simply do not want too close contact with their kin. If they have close and meaningful ties in later life it is generally with people in their own generation with whom they have more in common. There are plenty of individual exceptions to this generality, particularly where the mother–daughter relationship is concerned, but often it is the middle-aged daughter being emotionally dependent on her mother rather than the other way round. Peplau and her colleagues cite as many as eight studies which come to the conclusion that: 'Results from a growing number of studies suggest that social contact with friends and neighbours has greater impact on well-being than does contact with grown children or other relatives.'[20]

Peplau and her colleagues in their article 'Being Old and Living Alone' report that

> For those old people who are not currently married, living alone is a common pattern. But whereas stereotypes depict this as an unhappy necessity, old people are more likely to view living alone as a preferred lifestyle. For the unmarried, alternatives to living alone – living with children or relatives,

moving to a communal residence, sharing a home with somebody – are often unattractive. For example, although old people
want to be in geographical proximity to their children, they
do not want to share the same household. . . . Many an exas-
perated son or daughter worn out from continued pleading
with an ageing parent to 'Move in with us' or 'Let us arrange
for someone to move in with you' can testify that keeping
one's own place and keeping it to oneself is a matter of pride,
fiercely defended.[21]

In the study of members of the U3A reported in Chapter 3,
one 'Separated' man who preferred to live alone put it thus:
'I often feel it would be nice, if I live on my own, to be within
easy visiting distance of friends or family where I could drop
in occasionally. But at the same time I don't want to live in
my home town and be smothered by my extended family.'
(Case 118)

The common assumption that older people ought to be emotionally dependent on their children and younger relatives may
cause some older people to feel guilty. They may think that they
are in some way odd and lacking in natural affection because,
while remaining fond of their families, they prefer to be fond at
a distance, and to enjoy the blessings of solitude tempered with
as much, or as little, social intercourse as they choose. It is only
in more recent history, with the development of the Third Age
in fact, that older people are beginning to realize that they are
not to be considered as mere appendages to the younger generation. The misapprehension that many of the younger generation
have that their old parents *need* them for moral support is not
entirely disinterested. It is very convenient indeed to have grand-
parents, particularly grandmothers, to be at their beck and call
as unpaid household helps, childminders on occasion, and general
dogsbodies. Most older parents have earned their right to solitude
and they should insist that their children respect it.

Again, the plea that lone elderly parents should leave their
solitary homes and move into the family nest in their later years
has the immediate advantage for the children that the family
coffers are no longer being depleted by the running expenses of
a separate house, and that the sale of the house will bring in

some useful capital. Grandma or grandpa will then be living under the scrutiny, if not the actual control, of their children, and therefore less likely to do something 'silly', like taking on a toy-boy or a younger wife to upset the family applecart. Lone parents should be warned by the example of King Lear whose trust in the generous pretensions of Goneril and Regan proved to be disastrous.

Solitude as a therapeutic agent

In modern western society a high value is placed on 'together-ness' and it is assumed that if anyone is emotionally ill-attuned to life they are lacking in a proper degree of integration with their fellows. While no one would deny that love and friend-ship are important it may be that some individuals suffer from being too much at the mercy of the social group in which they happen to find themselves. Social groups vary very much as to their prevailing *Zeitgeist*, and sometimes the unhappy individual is one who is a square peg trying unsuccessfully to fit into a round hole. The modern practice of psychotherapy and counsel-ling that is so very popular often consists of trying to shave the corners off such square pegs to make them fit better, a practice that some contend is basically unsatisfactory.[22]

In Chapter 1 the case of E.M. Forster's fictional character Mrs Moore was considered, an English lady who went out to India in the time of the British Raj, and found herself utterly unable to adapt to the Anglo-Indian society and who retreated into a miserable existential loneliness. Although this is fiction it is an excellent example of the individual at odds with a particular social group to which she originally thought that she belonged.

Japanese society is perhaps even more committed to the pur-suit of 'togetherness' than that of the West, and an unusual form of therapy has developed there which is utterly different from the general conduct of what we understand by psychotherapy. Shoma Morita, a psychiatrist who was perhaps reacting from the practices of his colleagues and the growing influence of psycho-analytic ideas, developed a form of therapy that makes positive use of solitude as a means of attaining peace of mind and a satisfactory adjustment to society. Society makes demands on the individual, and the latter may become overwhelmed and confused by these demands; it is necessary, therefore, for the individual to have the opportunity of retreating into solitude and considering just what is the nature of his or her individuality, of evaluating the validity of these demands and how they can best be met or ignored.

Morita therapy, as it has come to be called, is somewhat complex, and is sometimes mistakenly believed to have arisen out of Zen philosophy. Karen Horney, when she visited Japan shortly before her death, had this idea but she was told by a prominent Morita therapist, Gen-yu Usa, who had once been a Zen monk himself, 'Morita therapy was never constructed on the ideas of Zen, nor was it explained in accordance with Zen . . . the founding of Morita therapy had nothing to do with Zen.[23] As a psychological treatment this form of therapy is not regarded as suitable for people suffering from hysterical types of disorders, but is applied to those who feel themselves to be disoriented and lonely in a society which is stressful, competitive and over-concerned with 'togetherness'.

A patient undergoing this type of therapy enters a hospital and is sequestered in a bare room with nothing but a mattress on the floor. Patients have nothing to read or occupy themselves with and can only leave the room to go to the bathroom. They may write their thoughts in a diary, and although they receive no visitors a therapist may look in on them occasionally, but not to talk to them, merely to look at their diaries and monitor their progress, for everyone is not left alone thus for the same number of days. This period of utter solitude generally lasts about a week, after which they are released into the second phase of the treatment, living in a work community but observing utter silence like a Trappist monk. Later, they progress to a third phase where they may assemble with others more sociably and attend lectures which are intended to instruct them in the the theory of Morita therapy, and how they may avoid *toraware* in the future, that is, the process by which social forces may confuse, dominate and disorient them.

It should be stressed that Morita therapy is utterly unlike psychoanalysis which seeks to *alter* the personality. It is made clear that we should come to terms with our own personality, to find out exactly what we are really like, to accept it and not to try to live by standards that are false for us. This is the same as was written up in the temple of Delphi in ancient Greece – 'Know thyself'. It is held that we can best attain such self-knowledge in solitude.

Morita therapy has never caught on in the West, although it has some parallels with the behaviour therapy movement which developed in the 1940s.[24] Anthony Storr, who has some leanings towards psychoanalysis, paradoxically has written the following about solitude having a therapeutic value that accords well with the ideas of the Morita therapists:

> The capacity to be alone is a valuable resource when changes of mental attitude are required. After major alterations in circumstances, fundamental reappraisal of the significance and meaning of existence may be needed. In a culture in which interpersonal relationships are generally considered to provide the answer to every form of distress, it is sometimes difficult to persuade well-meaning helpers that solitude can be as therapeutic as emotional support....
>
> Changes in attitude take time because our ways of thinking about life and ourselves so easily become habitual ... it has been realized that even elderly people are capable of change and innovation. Some people find it hard to adapt to any kind of change in circumstances; but this rigidity is more a characteristic of the obsessional personality than it is of being old.[25]

The rigidity of older people has been much over-stated, and alleged examples of it sometimes arise when younger people in positions of power, such as social workers, find that their elderly clients do not always agree with proposed changes that are supposed to be for their own good. Bernard Ineichen illustrates the first page of his book on dementia with a cartoon showing two social workers, one of them saying to the other, 'Senile dementia? Isn't that when elderly clients disagree with you about what's best for them?'[26] In fact, the whole experience of ageing means that the elderly person has to adapt to a great deal of change, and much of it may be very unwelcome.

Solitude and personal growth

There is no reason why we should ever cease to grow in moral stature whatever the length of our lives. Indeed, although some skills may deteriorate with ageing others may continually improve. It is reported that a student asked the great musician Pablo Casals why he practised every day, and he was told, 'Because I aim continually to improve.'

The pressure of ordinary life may be such that some people need periods of solitude in which to readjust their orientation to society. In his book *Alone* Admiral Richard Byrd described how he decided to seek the solitude of manning the Bolling Advance Weather Base alone during the Antarctic winter of 1934. He explained his decision thus:

> Now it is undeniably true that our civilization has evolved a marvellous system for safeguarding individual privacy, but those of us who must live in the limelight are outside its protection. Now I wanted something more than just privacy in the geographical sense. I wanted to sink roots into some replenishing philosophy. And so it occurred to me, as the situation surrounding Advanced Base evolved, that here was the opportunity. Out there on the South Polar barrier in cold and darkness as complete as that of the Pleistocene, I should have time to catch up, to study and think and listen to the phonograph; and maybe for seven months, remote from all but the simplest distractions, I should be able to live exactly as I chose.[27]

As time went on in his isolated state Byrd studied the changes in his psychological state just as had Edith Bone. After about two months he began to suffer from periods of depression in the evenings. He recorded in his diary:

> May 9th.
> I have been persistent in my efforts to eliminate the after-supper periods of depression. Until tonight my mood has been progressively better; now I am despondent again. Reason tells me that I have no right to be depressed. My progress in eliminating the indefinable irritants has been better than I expected.

I seem to be learning how to keep my thoughts and feelings on a more even keel, for I have not been sensible of undue anxiety.

May 11th.
I've been trying to analyse the effect of isolation in a man. As I said, it is difficult for me to put this into words. I can only feel the absence of certain things, the exaggeration of others. In civilization my necessarily gregarious life with its countless distractions and diversions had blinded me to how vitally important a role they really did play. I find that their sudden removal has been much more of a wrench than I had anticipated. As much as anything I miss being insulted now and then, which is probably the Virginian in me.

May 16th.
It's just a week since the last after-supper depression. I don't want to be over-confident, but I believe I have it licked.[28]

Byrd's initial period of depression in solitude which eventually yielded is fairly typical of living in solitude, as will be seen when we refer to Alexander Selkirk later on.

We may regard Byrd as fortunate in being able to obtain for himself such complete and utter solitude, as for the great majority of us the attainment of such a situation is impossible. However, I have before me part of a diary that was kept by an elderly man who was bereaved by the death of his wife, thus ending a marriage of long duration. This bereavement had upset him considerably and he even contemplated suicide as a way of ending the weariness, the apparent pointlessness of his future existence. This extract is of interest in that it shows that we do not need to have actual physical isolation to achieve what he calls 'the balm of solitariness'.

I went out after half-past ten having fallen asleep by the fire for a little but once awake felt very wakeful and faced the prospect of another sleepless night. I went out and walked along the river bank and there at the quayside were a number of young people chattering and drinking, and their voices

were loud and silly. There was a huge blaring volume of music from a ghetto-blaster, music that sounded just like the sort
of noise a machine would make. I felt unreasonably disturbed
by all this and tried to tell myself that – this is nothing to do
with you, so why should you worry? These people aren't really
happy or they wouldn't need such a blare of sound or to talk
so loudly themselves. I recalled Blake's poem:

> I wander thro' each charter'd street
> Near where the charter'd Thames doth flow,
> And mark in every face I meet
> Marks of weakness, marks of woe.

Lower down the river the towpath is over a rough common,
and it was there I looked up to the sky. It was simply mag-
nificent. It was dark enough here to see the stars, the great
blaze of them in a clear sky. My mood changed quite suddenly and I found myself in a state of what I can only describe
as awe, awe at the splendour of the universe in which I was
just a tiny speck, but privileged to be alive and to observe it.
What more could I ask? I was alone, quite, quite alone, and
glad to be alone. It did not matter that I had no-one to turn
to. An empty house and empty bed to return to, but somehow the balm of solitude, utter solitude, had spread over me
and I know that I will sleep tonight.[29]

What calms people and brings peace to them is very mysterious,
and in the case of this man living in a populous city, it was his
looking up to the sky and contemplating the very insignificance
of his own being in the immensity of the universe.

In Chapter 4 the case of Daniel Defoe's character Robinson
Crusoe was considered as an example of a man rendered utterly
alone and achingly lonely by being marooned on an island, but
the reality is even stranger. Defoe based his book largely on the
case of Alexander Selkirk, but if we turn to the contemporary
accounts of the 'rescue' of Selkirk given by Woodes Rogers and
Richard Steele we get a surprising narrative of how the loneliness which was first experienced turned to a very contented solitude
after a period of adjustment. Rogers writes:

He had with him his clothes and bedding, with a firelock, some powder, bullets and tobacco, a hatchet, knife, a kettle, a Bible, some practical pieces, and his mathematical instruments and books. He diverted and provided for himself as well as he could, but for the first eight months had much ado to bear up against melancholy, and the terror of being left alone in such a desolate place.[30]

Rogers mentions a period of eight months being the extent of Selkirk's melancholy, but the duration of his stay on the island was over four years.

Richard Steele put the period of Selkirk's melancholy as 18 months, and wrote:

The Necessities of Hunger and Thirst were his greatest Diversions from the Reflection on his lonely Condition. When these Appetites were satisfied, the Desire for Society was a strong Call upon him, and he appeared to himself least necessitous when he wanted everything; for the Supports of his Body were easily obtained, but the eager Longings for seeing again the Face of Man during the Interval of craving bodily Appetites, were hardly supportable. He grew dejected, languid and melancholy, scarce able to refrain from doing himself Violence, till by Degrees, by the Force of Reason and the frequent reading of the Scriptures, and turning his Thoughts upon the Study of Navigation, after the space of eighteen Months he grew thoroughly reconciled to his Condition. . . .

When the Ship which brought him off the Island came in he received them with great Indifference, with Relation to the Prospect of going off with them, but with great Satisfaction in an Opportunity to refresh and help them. The Man frequently bewailed his Return to the World, which could not, he said, with all its enjoyments, restore to him the Tranquility of his Solitude.[31]

This account is specially interesting in that it mentions that although Selkirk was not particularly happy to have his solitude ended, he still retained a normal benevolence towards other men in that he was pleased to help and refresh the sailors who took him off the island.

Earlier, the case of Richard Byrd's voluntary solitude was described and how he became depressed for a time, but recovered
later, actively appreciating his solitude, as did Alexander Selkirk.
Although Byrd's experiment ended in near-disaster as he became
poisoned with fumes from his defective stove, four years after
his experience he wrote:

> I did take away something that I had not fully possessed be-
> fore: appreciation of the sheer beauty amd miracle of being
> alive, and a humble set of values. . . . Civilization has not al-
> tered my ideas. I live more simply now, and with more peace.[32]

Considering the whole question of solitude, we may come to
appreciate that it is not something to be dreaded as a most unwelcome concomitant of later life. Ageing is an adventure, and
if an increased degree of solitude is to be our lot, perhaps we
should regard it as a positive bonus which will add to the rich-
ness of our lives.

6
Overcoming Loneliness

In Chapter 3 the principal reasons for loneliness in later life were presented, both those gathered from published sources and those volunteered by the members of the University of the Third Age who took part in the survey. It should be remembered that the latter were a rather special group of people, and that the fact of their having joined such an organization may have resulted in their being less lonely than the generality of the population in their age-group. However, the best way to proceed seems first to consider the main reasons for loneliness that emerged from this study.

Results from the survey

One of the limitations of gathering information by means of a questionnaire such as was used in this survey is that, in order to ensure a relatively high rate of return, only a small number of questions can be asked. Experience in research with questionnaires has shown that the more detail that is asked the more people will be reluctant to fill in and send them back. Thus in asking about marital status it would have been very useful to know how long widows and widowers have been bereaved, as this may make a material difference to their present degree of loneliness. As it is, the material that is available will include those who have been bereaved recently and those who have been rendered single for quite a long time. The same applies for those who are in the divorced/separated category.

Bereavement is clearly the principal cause of loneliness for the elderly, and generally it is due to the death of a spouse. The break-up of a marriage may also cause loneliness, but for the generations which are now in the Third Age divorce and separation are not so common as with younger people, and in some cases the ending of an unsatisfactory marriage in later life may bring peace and contentment rather than loneliness. In planning the questionnaire it was not thought advisable to ask those who were officially 'single' whether they had a close and intimate relationship with someone; in fact one widowed lady (Case 45) volunteered the information that she has such a relationship with a widower, and added 'If it were not for him my answers to loneliness questions would be very different. What I remember of widowhood experience before I moved here and met my widower friend was that I was utterly dismal and lonely, and that family and friends only partly filled the void.'

In the Survey (Table 3.7) the item which was most frequently endorsed by women as being a reason for loneliness was 'A Loneliness which represents a desire to continue a relationship with someone who is no longer available'. This might conceivably refer to desertion by husbands, or loss of a very dear friend or relative, but it seems most likely that it generally referred to marital bereavement. The items which were next most frequently endorsed by women in this context were G (Loneliness for a former lifestyle), C (Loneliness due to absence of anyone you care for) and D (Loneliness for a personal relationship similar in depth to a lost relationship). These items obviously refer to bereavement as well as to other losses in later life.

As far as men are concerned it is unfortunate that only 17 out of the 62 men contributed to this Table; two probable reasons may have accounted for this: first, men were, on the whole, less lonely, and second, men tend to be less communicative about themselves. As it is, any conclusions that may be drawn must be very tentative. The item most frequently endorsed by them was 'G. Loneliness for a former lifestyle' which, of course, might refer to the lifestyle of a married life, as did the next most frequently endorsed, 'C. Loneliness due to absence of anyone you care for'. Further results from the survey will be discussed later.

Bereavement

Bereavement is a topic which has attracted considerable attention from the caring professions and there are a number of societies which offer help to the bereaved, such as those given in Appendix A. Anyone who is suffering from the sad effects of the death of someone they love should consider seeking help and comfort from those professionals and organizations who are qualified to give it. They may also wish to read publications on the subject, and some of those directed to the lay public are listed in Appendix B. It is recognized that of all the experiences that are seriously traumatic even for the most well-balanced people, bereavement is probably the worst. There is no merit in trying to 'keep a stiff upper lip'; grief must be acknowledged and a period of mourning gone through to enable the bereaved man or woman eventually to readjust their lives.

It is unfortunate that one of the great taboos in the western world, and perhaps especially in Britain, is acknowledging the reality of death in a matter-of-fact way. It sometimes leads to a most unhappy position of the bereaved person being almost ostracized by the friends whom he or she expected to be supportive. Such friends do not mean to be unkind, it is just that they are so overwhelmingly embarrassed by the subject of death that they do not know how to relate to the bereaved one. According to Alex Comfort:

> There is no way of *dealing* with bereavement so as to make it painless. Neither the British technique of pretending that death didn't happen nor the American mortician-promoted technique of cosmetics, exploitation and open coffins works. Both tend only to limit the normal expression of normal emotions of grief, rage and despair which surface a bit later as depression or illness. Bereaved people tend to be quite unexpectedly boycotted by friends who don't know how to handle death – and stay away.[1]

Bereavement is one of the most serious problems of later life causing people to be lonely: this stems from a number of attendant factors. Not only does the bereaved person miss the presence of the loved one, but such a loss leads to a falling-off of the social contacts that were dependent on the presence of the partner. For example, a wife may have had many contacts through her husband's friends, and when he is dead these contacts may no longer be available to her. Indeed, her husband's friends may actively shun her, feeling guilty all the time, but not knowing how to cope with her on her own with, as it were, the aura of death about her.

Most people are living in couples when they reach the age of retirement and although they may not be very dependent on each other in the relationship, when it is dissolved through the death of one of them, the survivor may feel far more bereft than was expected. On logical grounds it might be supposed that the death of an elderly person might be less of a tragedy than the death of a young or middle-aged relative or friend because the latter might have a considerable length of life to fulfil. However, illogical as it may seem, the death of an elderly spouse or friend may be even more traumatic. In later life most couples (although they may not regard themselves as being particularly dependent on one another, or even think of themselves as a 'couple'), have come to be like two beams in a building propped against each other, and, if one is removed, the whole structure collapses. The surviving partner may have to cope with things that he or she never thought about before, and although it sounds shamefully mundane, it is often trivial things that assume an overwhelming and disproportionate importance. Widowers tend to be harder-hit than widows, and there is a raised incidence of death among men in the year following bereavement. The bereaved person may feel ashamed of some aspects of his or her deprivation; Wendy and Sally Greengross write of one aspect of bereavement that many people do not like to admit to, especially when they are of quite an advanced age:

> The pain of this overwhelming loss is often made worse because the surviving partner quite unexpectedly finds that he or she has sexual feelings that demand satisfaction. Some of the need may be for touching, comfort or caring, but some people still have strong physical sexual desires. These create confusion for those for whom it is still a struggle to find

emotional stability, and the lack of open discussion arouses fear that no-one else has such feelings or desires. The conspiracy of silence that society imposes on this topic often prevents widows and widowers seeking appropriate advice and reassurance.[2]

Sometimes, to the horror of the bereaved person, he or she experiences an *increase* in sexual desire and yearning, as reported by Swigar and his colleagues,[3] but this is simply a perfectly understandable and primitive desire for physical contact to alleviate loneliness, just as a lonely child wants to be cuddled.

There is a sense of loss that is difficult to describe; some writers have referred to it as 'lack of a confiding relationship' and it is important in the generation of a depressed state. Women are rather better than men at forming 'confiding relationships' in our society, and, as mentioned above, in a large section of Christian society in Britain men may make their depressed state worse by endeavouring to 'keep a stiff upper lip' after bereavement. We may contrast this with the institution of *shivah* in the Jewish religion which prescribes that for seven days there is a very definite effort to support the chief victims of bereavement. During this period prayers are said for the dead and the mourners are expected to spend much of their time talking to visitors about the dead person.[4]

Murray Parkes has criticized the conduct of this Jewish rite in contemporary Britain:

> My impression from talking to a number of intelligent middle-class Jews is less favourable than Gorer's. They have pointed out that while it is true that the *shivah* still serves its traditional function of drawing the family together at a time of bereavement there is a tendency for it to be used as a distraction from grief rather than an occasion for its expression. Conversation with the bereaved person often takes the form of neutral chat and the expression of overt emotion is avoided as it is in other 'public' situations. The 'successful' mourner is thought to be one who shows a proper control of his feelings on all occasions.[5]

How true Murray Parkes's impressions are must depend upon individual cases, but it will be noted in Appendix A that the Jewish Bereavement Counselling Service goes to considerable trouble to see that all bereaved people who need help according to the institution of *shivah* will receive it.

Various cultures over the world have different social customs designed to overcome the shock of bereavement and to help the bereaved overcome their loss. Anthropologists have recorded very many of these customs and they generally involve a rather public mourning and an emphasis on, and even exaggeration of, the grief that the death has caused. For example, mourning rites in some parts of India involve the hiring of professional mourners who express grief in a very histrionic manner, even though they have not known the deceased. The object of these various mourning ceremonies is to get the surviving spouse and immediate family to accept that the death is not just a private loss but a loss to the whole community, and they are not alone in a private sorrow.

The dreadful modern taboo on the mention and acknowledgement of death in many western countries leads to an unexpected withdrawal of social support making the widow or widower feel additionally alone. This is a sociogenic component that makes the loss specially painful. The process of mourning must be got through and this may involve facing some unpalatable facts that arouse guilt and hence a self-punishing depression. It is perfectly natural in some circumstances to feel *anger* against the deceased spouse for their 'desertion' – 'If only the fool hadn't drunk so much he wouldn't have died!' 'If she'd had the sense to take *my* advice about her health she'd still be here!' It is also natural to feel a sense of *relief* that the loved one has died, especially when there has been a long illness that has been stressful for the carer. These feelings about one's own selfishness and human weakness should be frankly acknowledged during the mourning process, as well as the remorse that can be summed up as, 'If only I'd been a more caring wife/husband maybe he/she would be alive today.' Depression generally has a component of self-punishing guilt that can only be got rid of by consciously admitting reality rather than shrouding the whole matter of bereavement in conventional platitudes.

Many writers on the subject of mourning have divided its progress and recovery from it into a number of stages. Murray Parkes has outlined five stages – alarm, searching, mitigation, anger and guilt, gaining a new identity.[6] Others have described three stages – shock and disbelief, developing awareness, and resolution.[7] Cook and Phillips describe the process of mourning with its attendant loneliness in terms of the three stages outlined below.[8]

1. The immediate impact of loss

The immediate impact of bereavement is much the same whether it has been a sudden death or one which has been anticipated for some time when it is known that someone is dying. There may often be a strange disjunction between what is known to be true and a half-belief that reality is otherwise. The widower may know quite well that his wife is dead but finds himself searching for her in a crowd because momentarily he has seemed to see her. He may wake in the morning and be surprised not to find her next to him.

In the shock occasioned by a recent death the bereaved person may experience a certain numbness so that, even when in the presence of friends and engaging in conversation, there is nevertheless a sense of appalling loneliness and an inability properly to comprehend what is being said. The loneliness is not dispelled by their company and the sufferer may even wish that they would go away and leave him/her in peace. All that has been said in the previous chapter concerning solitude is relevant here. The bereaved one may need a measure of solitude to work through the all the unpleasant feelings that have been mentioned earlier. Extreme pessimism characterizes this first stage of mourning. Intelligent people will realize that in the course of time they will recover from the misery of bereavement and be on an even keel again because they have seen this happen to other people, but it may seem at the time that this is quite impossible and that there is no light at the end of the tunnel.

2. The process of change and developing awareness

If a bereaved person, say a widow, has responsibilities she cannot avoid and she has to go on busying herself as usual, she may indeed repress her feelings of loss and despair and outwardly appear not to suffer much. Friends may comment on how well she is coping, and she herself may think that this busyness may overcome her sense of devastation. To some extent this is an illusion; her non-expression of emotion may prolong the process of coming to terms with her new and altered state; life will never be quite the same again and she might as well admit it. Being denied solitude in which to readjust her life may do her no good at all.

People may use various techniques temporarily to overcome their sense of loss, and one of them is to idealize the deceased, and this may be especially true when there was indeed a good deal of discord in the relationship. Only the good aspects of the deceased are admitted and any criticism of him provokes anger in a widow who sets out to idealize a dead husband. Japanese folk-tales present a theme of the ghosts of dead members of a family being spiteful against the living and having the power to injure them if they do not constantly make a show of placating them, and it may be that in idealizing our departed loved ones we are moved by a superstitious fear. It is best to try to work towards a realistic appreciation of the fact that in all relationships there may be an element of conflict, and that there were faults on both sides. When a more realistic image of the deceased takes over, the bereaved begin to come to terms with their lonely state, and evaluate it for what it is, and look forward constructively to the future.

It is easy to confuse loneliness with depression. The latter term is used to indicate a persistent lowering of mood that is more serious than just being down in the dumps, thoroughly fed up or temporarily being afflicted by 'the black dog'. Depression implies suffering from a definite illness: traumatic events (such as bereavement) can cause an actual alteration in the chemistry of the brain, and while this physiological condition persists, one's normal adjustment to life is impossible. Psychiatrists used to refer to 'reactive' as contrasted with 'endogenous' depression, the latter term referring to a more serious condition that has arisen from some basic physiological disorder, in contrast to just a strong reaction to ongoing circumstances. Nowadays it is more usual to refer to the milder reactive form as 'dysthymia' and the more serious form as 'major depression'. It is not unusual for the experience of bereavement to bring on a period of dysthymia, where the sufferer may feel very 'lonely' even though surrounded by friends and family in a close supportive network. People in this condition frequently go to their doctor who will prescribe some form of anti-depressant drug, but unfortunately with some people these drugs have no effect on the depression. This is not the fault of the doctor, for it is an area of psychological and medical debate and a great deal more needs to be found out about the physiology of this state. Apart from drug treatment there are some forms of psychological therapy which are used in the treatment of depression.[9] Fortunately states of minor depression tend to be self-limiting, and time alone works a cure.

3. Rebuilding a life

The time it takes to recover from the experience of bereavement varies very greatly according to the personality of the bereaved person, the circumstances in which they are placed, and what the deceased has meant to them. Obviously the loved one cannot be replaced entirely, as every individual and every relationship is unique, but there is no reason why those who found fulfilment in an earlier marriage, or such a relationship, should not find happiness in the future with somebody else if they should be so fortunate as to meet someone suitable. It may be that a future partner and a future relationship may be very different from what was fulfilling in the past, for we change as we age and what was necessary in the past may no longer apply. Elsewhere I have published a collection of autobiographical accounts written by people who have formed new love-relationships after the age of 60, and these accounts are very instructive.[10] One lady who contributed to the book ended her account thus:

> My conclusion is that we cannot afford to deny love in later life, whatever its imperfections. It may be of a kind we should never have considered in earlier, more idealistic days, but now we must enjoy what we may while we may.

Another contributor who had been a wife and mother met another woman in later life with whom she fell in love and formed a very satisfactory relationship with her. She said of her relationship:

> I know that all this could end for a variety of reasons, just as relationships did when I was young, and I also know that I would be devastated and heartbroken, but it will have been a worthwhile experience. I would never say no to love. It is so precious that I think that we should open ourselves up to it wherever we find it, no matter what the source. Somehow we expand with love and life seems more possible. I know that Marie feels the same and we would be lost without each other.

Of course everyone who is lonely through bereavement does not necessarily want to find a new relationship to replace the lost one, and some people have quite other plans for their future and aim to find satisfactions in life that do not depend on close personal relationships. This is just as well for women because of the gross numerical disparity between the sexes in later life, there being about twice as many women as men by the age of 75. One elderly commentator on the research that led to my earlier book wrote as follows:

> Over three million of us are widows. I feel that most of us have spent most of our lives in a close relationship and would prefer to live this way. Sometimes a feeling of loyalty to our lost partners, or lack of confidence in ourselves, leads to a withdrawal from society, and I am all in favour of any measure that helps to bring us back to life. . . .
>
> Of course there are some men around, even after eliminating those who are already married or are seriously unfit. But there are still problems ahead. Gold-diggers and con-men have a field day, but can usually be spotted in time. More tricky is the 'hot meals and slippers' syndrome; those men who are really looking for an unpaid nurse/housekeeper for their 'old age'. Fortunately for them there are many women who relish such a role, but at least both sides should know where they stand.
>
> I am in favour of a realistic approach. Of course one can fall in love at any age and have that love reciprocated. Such people are extremely fortunate and I wish them all the best.

But surely it is a waste of the precious days left to us not to realize the odds against finding another partner, and face up to life alone. The greatest advantage is freedom, freedom to come and go as you please, freedom from domestic routine, freedom to experiment and try out new lifestyles, perhaps discover new aspects of yourself you never knew existed. The greatest disadvantage is loneliness. No-one to talk things over with, to give encouragement when the going is tough. If you have been used to life in couples it is hard to find you no longer fit in.[11]

This elderly lady describes the problems of widowhood very well, but she seems to take for granted the conventions in which she grew up. She does not appear to realize that times have changed and that there are already a certain number of widows in the Third Age who certainly enjoy all the advantages that she describes but who who are not bound by the old conventions. As demonstrated in my earlier book, some widows live on their own but have lovers who also live singly but who certainly supply much, if not all, the human warmth and comfort that their husbands used to give. That there are not enough men to go round is certainly true, but as exemplified in the case of the lady quoted earlier, who fell in love with another woman and had her love reciprocated, more women in later life are now forming mutually supportive pairs with or without an erotic component to their relationship.

One factor that may deter some bereaved people from remarriage or the formation of a new emotional relationship is a feeling that it would be 'disloyal' to the deceased partner. Such a feeling is often connected with the mechanism of 'enshrinement'; the unrealistic overvaluing of the lost loved one that has been referred to in Stage 2 of mourning above. Certainly a sufficiently long period of mourning is necessary before any such step as remarriage can be reasonably entertained, but if mourning is too prolonged it becomes pathological. Butler and Lewis give the following advice:

After the death of a partner it is often very difficult for the man or woman who has been widowed to look ahead to a new partner without feelings of guilt or disloyalty to the memory of the dead one. In *enshrinement*, the survivor keeps things just as they were when the loved one was alive and spends his or her energy revering the memory of the dead person, surrounded by photographs and rooms maintained intact. The survivor believes that to live fully is a betrayal of love or loyalty for the dead. This survival guilt and fear of infidelity leads to emotional stagnation and stands in the way of forming new relationships. Once the period of mourning is over and the initial shock and grief have abated, you owe it to yourself to become realistic about your need to have a new life of your own. This means the appropriate preservation of your memories without excessive dwelling in the past. The usual cure for enshrinement is to take an active role in getting life moving again. This is an act of will and determination. It can happen only if the individual decides to make it happen. Removing from sight the personal possessions of the deceased will help. It may also be necessary to put away obvious marriage symbols, such as a wedding ring. It is not a betrayal of a past marriage to accept the present and build a future.[12]

As I have discussed elsewhere,[13] the children of someone who has been rendered single in later life by bereavement may have a vested economic interest in discouraging their parent from remarriage, or their objection may simply be based on neurotic jealousy. One of the possible ploys is to accuse their parent of disloyalty to the dead spouse. Such children are reprehensibly selfish; they would rather see their parent single and lonely than settling into a new life with a loving and caring partner.

The divorced and separated

Loneliness resulting from the break-up of a marriage or a long-term love relationship can be equally as severe as that which follows bereavement. Added to the same factors associated with bereavement, a divorce or separation frequently produces very definite anger against the partner who is viewed as 'disloyal' or 'uncaring', and there may be a measure of personal guilt for having 'failed' to be a successful husband or wife. At least bereavement does not normally engender such bitter feelings. The damage which has been done to the self-image may take a long time before it is repaired, and the formation of a new and successful relationship later on may be the best remedy, and here the person rendered single will have no qualms about disloyalty to the former partner.

Turning to the results of the survey, of the 39 currently unmarried women, 25 were widowed and five divorced/separated. Of the nine unmarried men, four were widowed and four divorced/separated. This seems a rather low figure in the latter category for the women and it is possible that a number of those who had been through the unpleasant experience of divorce or desertion did not complete and return the questionnaires because the subject was painful to them. Of course, as pointed out at the beginning of this chapter, the important detail of how long ago the traumatic event had taken place was not asked. For some it may have taken place quite long ago and they have had time to readjust their lives and so were no longer lonely. The address of the National Council for the Divorced and Separated is given in Appendix A, and some books giving advice to the divorced/separated are listed in Appendix B.

Loneliness due to loss of status

In the list of situations causing loneliness presented in Chapter 3 (Table 3.7), item H, 'Loss of status', did not attract much endorsement, but it was claimed significantly more often by men than women. A man's status in life, particularly for those who are now in the Third Age, has depended very much on his work, but this has been true for women to a lesser degree. Retirement hits some men very hard, especially if they have occupied positions of importance and responsibility, and have let their work become too paramount in their life to the exclusion of domestic and leisure interests.

In the academic world it is not uncommon for professors and other senior staff to be granted the continued use of an office in their department after they have been retired by reason of their age. This is more or less an act of charity, and the elderly academic strives, not always successfully, to keep up with his or her subject, and may be conscious of the fact that younger colleagues regard them with a certain condescension as they themselves forge ahead in a rapidly developing subject. Endeavouring to hang on in a place where one is no longer a 'big shot' can engender a sense of loneliness when the erstwhile incumbent of an important post has not developed interests outside work.

This sort of loneliness may be engendered in the world of business, or sport, or in fact in any field where practices are in a state of development and flux, and the young are introducing totally new methods. The old have the advantage of long experience and accumulated knowledge, but what if innovations in practice have rendered this knowledge irrelevant?

Arthur Miller's play *Death of a Salesman*[14] is a very insightful and poignant study of Willy Loman, a man who has let his job as a salesman take over his whole life. He had been a highly successful salesman in the past, but he did not appreciate that not only was he changing with age, but the world of commerce was changing too, and that his skills were no longer appropriate. He was eventually thrown aside quite ruthlessly by the firm to which he had been devoted for almost the whole of his working life, and having neglected his family and other interests for many years he had no other resources. This led to utter loneliness, mental breakdown and eventual suicide.

The psychiatrist Carl Rogers has propounded a theory embracing what he calls the 'phenomenological perspective'.[15] He assumes that society pressurizes the individual to act in certain socially approved ways, and this may lead to a discrepancy between one's true 'inner self' and the self manifested to others. Simply performing a social role does not maintain and develop this 'inner self'. Rogers writes:

> Loneliness . . . is sharpest and most poignant in the individual who has, for one reason or another, found himself standing, without some of his customary defenses, a vulnerable, frightened, lonely but *real* self, sure of rejection in a judgemental world.[16]

According to Rogers those who do not trust their real selves to command the respect and approval of others use their social role as a shield to protect them in a world they perceive as hostile. Thus a man who has managed to lead a fairly successful life in the role of a sergeant-major, once he has retired to civilian life may find that other people do not give him the respect to which he is accustomed, and if his wife dies and his children are grown up and far from home he may be a particularly lonely and vulnerable widower.

Carl Rogers' theories were, of course, developed in the course of his work with emotionally disturbed patients, and are therefore not entirely applicable to ordinary, emotionally stable people. However, we may all learn something from this point of view. In preparing for retirement, and the various changes that come with age, we should be aware of all possibilities and what strategies we should adopt to avoid the rocks ahead. In discussing types of loneliness in Chapter 1 the variety identified as 'fantasy loneliness' was mentioned, and those who lose their prized status in later life are especially likely to develop this. Fortunately those with natural managerial skills and experience are by no means left resourceless in later life. In a previous book[17] I referred to two contrasted types in the Third Age – the 'task-oriented person' and the 'lotos-eater'. I addressed my readers thus:

> Are you a 'task-oriented person' who loves to buzz around organizing things, sitting on committees, and setting the world to rights? Or are you the opposite, a 'lotos-eater' who prefers to lie in the sun, who spends hours and hours just reading novels, listening to music and chatting with friends? These two types of individuals will adjust to retirement in different ways, but either may be equally fulfilled providing they adopt a life-style that suits them. They will be ill-advised if they try to follow one particular course of action because it is recommended by a 'specialist'.[18]

In studying the University of the Third Age I have seen how these two types of retired people complement one another: the 'lotos-eaters' perform a very valuable service in providing the other type with the opportunity to exercise their managerial skills unpaid. The whole organization depends to a great extent on those who used to hold important positions in business, the academic world, finance, and so on. Some of them continue to be as busy as they ever were before they retired and they have the satisfaction of seeing the 'lotos-eaters' happily and gratefully benefiting from their skills. All the classes, of which there is a huge variety, are given *gratis* by members, and this gives great scope for ex-teachers and other academics to continue to exercise their profession up to any age.

How then do we cope with the loneliness due to loss of status that so frequently follows retirement? The answer lies in not 'retiring' in the usual sense unless we are happy as a 'lotos-eater' as some people are.

The following quotation comes from the great Spanish musician Pablo Casals when he was interviewed at the age of 93:

> On my last birthday I was ninety-three years old. That is not young of course. . . . But age is a relative matter. If you continue to work and to absorb the beauty of the world about you, you find that age does not necessarily mean getting old. At least, not in the ordinary sense. I feel many things more intensely than before, and for me life grows more fascinating . . .
>
> Work helps prevent one from getting old. I, for one, cannot dream of retiring. Not now or ever. Retire, the word is alien and the idea inconceivable to me. My work is my life. I cannot think of one without the other. To 'retire' means to me to begin to die. The man who works and is never bored is never old. Work and interest in worthwhile things are the best remedy for age. Each day I am reborn. Each day I must begin again.[19]

It may be objected that very few of us have the great talent of Casals, but that is really irrelevant: each of us has our own unique interests and talents that can be developed for the whole of our lives. Obviously they may have to change in later life to some degree: the athlete no longer has his physical skill and the singer may lose his voice, but ageing is an adventure that presents new challenges and calls for the development of new interests and skills. The work we do in later life may be an extension and modification of what we used to do or something quite fresh, but whether our later interests and occupations take the form of gardening, studying for an Open University degree, dressmaking, taking up oil-painting, working for a charity, joining a campaign for some cause, extending our foreign travel, participating actively in one of the many special-interest clubs that are available – that is up to us. We do not need any very special talent and to be acknowledged by the world; it is our life to live as we please.

Opinions from the survey

In Chapter 3 quite a number of the opinions of people who took part in the survey were quoted, and mainly they were volunteered by those explaining why, currently or in the past, they were lonely in some degree, or from others who feared that they might be lonely in the future if certain events took place. It is not proposed to quote any more of these, but as we have a great deal of material offered from others who are *not* lonely and are glad to offer their opinions on the subject it may be worthwhile quoting at least one. This is not offered as being specially insightful or from anyone who has made a special study of the problems of ageing and loneliness. Plenty of 'authorities' have been quoted in this book, and the following is from someone who claims no special authority and her contribution is of value in that it is very typical of what many women in her position have said, so rather than quote quite a number of opinions hers is offered because of its typicality. She is a married woman in the 71–75 age bracket, in good health and apparently active in the U3A.

> I think that in the Third Age people need some time to be on their own, but are not lonely. Even with a partner they need to retain their own individuality, for they may be on their own one day. I believe very firmly that 'life is what you make it'; by giving we receive. If one is shy or a reticent type of person it is more difficult to mix with others, but the effort has to be made, or loneliness, as distinct from being alone, is the result. Of course if one's health is poor and one is unable to get about things are difficult, and then one has to rely on others for help and encouragement. One must

always try to be interested and interesting – keep one's mind ac-
tive even if the flesh becomes weak.

The main causes of loneliness in later life I feel are due to lack
of friends and friendship. Mental resilience is required in greater
measure as one ages in order to overcome illness, and depression
resulting from ill-health. It is necessary to adopt a positive ap-
proach to life whatever one's circumstances may be – to look forward
rather than back. Age is just a number on one's birth certificate
and if you think young you'll feel young. Always have the next
day planned (take a day at a time). Never wake up thinking 'What
am I going to do today?' I have found that the majority of people
I have met in the U3A are very busy and positive, and the U3A
has opened up a new vista of life for them.

One should try to adopt an optimistic approach to whatever comes
along – also to life generally. If we sense that anyone is lonely we
must help them by whatever means we can. A lot of people who
live alone like to be alone and are by no means lonely – so we
should respect their wish for privacy.

I hope my comments will have been of some use.

It is of interest to note what this lady does not mention in
her prescription for a well-adjusted life in its later decades. She
does not mention relations with family, nor does she mention
religion; she is entirely peer-oriented and this is typical of most
of the others who have have responded to the request for opinions
in the survey, and indeed, comparable researches have found similar
results. Peplau and her colleagues reviewed a number of studies
and report:

These studies suggest an explanation for the common finding
(cited above) that contact with kin does not reduce loneliness
and enhance psychological well-being among older adults, while
contact with peers often does. A study by Arling suggests that
older adults engage in rather different activities with friends
than with grown children, and that relations with friends may
involve more reciprocal exchanges of assistance than those
with kin. Arling also found that the more people with whom
an old person engaged in *reciprocal* exchanges, the lower the
person scored on a measure of 'lonely dissatisfaction'. Further

research is needed to clarify the reasons for the differential effects of relations with friends versus kin . . . On the basis of studies reviewed here, we would expect social relationships with kin (whether with grown child, sibling or spouse) to be perceived as satisfying to the extent that they are character- ized by the positive qualities found in the relationships with friends and confidants.[20]

It is not suggested here that the lady whose opinions are re- ported above, and the many women like her, are at all indifferent to family ties, and some of them do indeed attend church, but more as the habit of a lifetime rather than any special religios- ity. As Murray Parkes observed in relation to his researches with widows:

In general I have the impression, and it is no more than an impression, that several of the more religious widows were insecure women who tried to find, in their relationship to God, the same kind of support they had sought from their husbands. Since such women tend to do badly after bereave- ment it is not surprising that 'faith in God' and 'regular church attendance' were not necessarily related to good outcome fol- lowing bereavement.[21]

It is difficult to get any reliable information about what the attitudes of the grandparent generation would have been, say, 50 years ago, but what clues we have suggest that they would have been very different. What we now have is a generation of older people in the Third Age who are sometimes referred to as the 'New Old'; they have quietly participated in what Starr and Weiner have termed 'The revolution of the old'.[22] Undoubtedly this revolution has affected women more than men and it is partly due to the postwar success of feminism, but I do not imagine that the majority of such women would regard themselves as 'feminists', nor would many of the modern radical feminists re- gard them as such. Many of the latter have distanced themselves from any concern with older women and the possible sources of loneliness and marginalization in society that affect them. Assister and Carol write of how '52 years of pauseless campaign' – the

Women's Suffrage Movement – '[c]ame to be known to our own generation, by vague derisive references to prim little old ladies who, briefly, shook their umbrellas and squawked for a bit, until the men rolled their eyes and magnanimously offered female citizens the vote'.[23] This denigration of the older feminists such as Emmeline Pankhurst by the young radical feminists has understandably meant that, according to the above authors:

> Today, some of the brightest women we meet, the ones who seem to have the most developed feminist understanding of sexual issues, are women who refuse to call themselves feminists at all, so loath are they to be associated with a movement that has become so steeped in an ideology of punitive and restrictive attitudes toward women. These are the women who insist on the authority to authenticate themselves, rather than letting others, once again, become the 'experts' on their lives. These are the women who actually argue the issues, who criticise real sexism where it is otherwise accepted, rather than using old rhetoric to throw darts at trivial icons.[24]

The time has come to identify what may be called a Third Age feminism, a feminism that is lived rather than proclaimed so stridently that one must doubt the emotional stability of its propagandists. The ethos of this Third Age feminism finds expression in a number of popular journals such as *Active Life*, *Choice*, and *Saga Magazine*, magazines that present an entirely new image of people in later life, an image that would have been indignantly rejected by the majority of the grandparent generation of even 30 years ago. An excellent picture of Third Age feminists is conveyed in a book published fairly recently entitled *Growing Old Disgracefully*.[25] Here the term 'disgraceful' is used not in its usual derogatory sense but part-humorously to distinguish it from the old concept that women should study how to 'grow old gracefully' with all its associations with resignation, lavender and lace. The book is authored by six women who came together to combat the threat of loneliness in later life in response to the fact that there are various disabilities looming ahead, not the least of which is the increasing shortage of available men. These women are not man-haters as are many of the more extreme of the younger radical feminists, but like Chaucer's Wife of Bath, are open to all new experience should fate throw it their way:

> Blessed be God that I have wedded fyve!
> Welcome the sixte, whan that evere he shal.
> For sothe, I wal not kepe me chaast in al;
> When myn housbond is fro the world ygon,
> Som Cristen man shal wedde me anon.[26]

This group of six women are certainly not dependent on men; they are realists and have come together for friendship and mutual aid. They have founded *The Growing Old Disgracefully Network* and offer to put older women in touch with like-minded women in their area; the address of this organization is given in Appendix A.

The 62 men in the survey did not offer much advice about the overcoming of loneliness in later life. As already stated, the great majority of them were living contentedly with their wives, and because of the limitations of the questionnaire it was not clear how many of them had been bereaved, or otherwise rendered single, and had remarried. The question of the loneliness of some of them, either married or single, due to a reduction in their status, has already been discussed. It is fair to say that the factors that produce loneliness in women in later life also apply to men, but that men in general are favoured by the demographic structure of the population as regards the relative numbers of the two sexes in the later decades of life.

Notes

Introduction

1 N. Bradburn, *The Structure of Psychological Wellbeing*, Chicago: Aldine, 1969.
2 Samuel Pepys, *The Diary of Samuel Pepys*, London, Macmillan and Co., 1906, p. 166.
3 J. de Jong-Gierveld and J. Raadfelders, 'Types of Loneliness', in L.A. Peplau and D. Perlman, *Loneliness: a Sourcebook of Current Theory, Research and Therapy*, Chichester: J. Wiley and Sons, 1982.
4 M.B. Parlee, 'The Friendship Bond: PT's Survey Report on Friendship in America', *Psychology Today*, October 1979.
5 J. Tunstall, *Old and Alone*, London: Routledge and Kegan Paul, 1967.
6 David Lodge, *Therapy*, London: Secker and Warburg, 1995.
7 E. Cumming and W. Henry, *Growing Old: the Process of Disengagement*, New York: Basic Books, 1961.
8 H.B. Gibson, *A Little of What You Fancy Does You Good: Your Health in Later Life*, London: Third Age Press, 1997.
9 Peter Laslett, *A Fresh Map of Life*, Second Edition, London: Macmillan, 1996.

1 What is Loneliness?

1 M.B. Parlee, 'The Friendship Bond: PT's Survey Report on Friendship in America', *Psychology Today*, October 1979.
2 Bertrand Russell, *The Autobiography of Bertrand Russell*, Vol. 1, London: Allen & Unwin, 1967, p. 13.
3 *Ibid.*, p. 146.
4 Rubin Gotesky, 'Aloneness, Loneliness, Isolation, Solitude', in J.M. Edie (ed.), *An Invitation to Phenomenology*, Chicago: Quadrangle Books, 1965.
5 Ronald Rolheiser, *The Restless Heart*, London: Hodder and Stoughton, 1979.
6 L.A. Peplau, T.K. Bikson, K.S. Brook *et al.*, 'Being Old and Living Alone', in L.A. Peplau and D. Perlman (eds.), *Loneliness: A Sourcebook of Current Theory, Research and Therapy*, Chichester: Wiley, 1982.
7 Jonathan Swift, *Gulliver's Travels*, Harmondsworth: Penguin Books, 1967, p. 258.
8 Rubin Gotesky, *op. cit.*, p. 236.
9 Peter Ackroyd, *Dickens*, London: Minerva, 1991, p. 548.
10 Ronald Rolheiser, *op. cit.*, p. 71.

11 W.F.R. Stewart, *Loneliness the Other Handicap: Loneliness – Its Causes among Physically Disabled People*, Amberley: Disabilities Study Unit, 1982.
12 Alex Comfort, *A Good Age*, Revised Edition, London: Pan Books, 1990, p. 116.
13 See J.A. Muir Gray, E.J. Bassey and A. Young, 'The Risks of Inactivity', in J.A. Muir Gray (ed.), *Prevention of Disease in the Elderly*, Churchill Livingstone, 1985.
14 Oliver Wendell Holmes, quoted by Rubin Gotesky, *op. cit.*, p. 228.
15 E.R. Bates (ed.), 'The Book of Genesis' in *The Bible Designed to be Read as Literature*, London: Heinemann [no date given].
16 Plato, *The Symposium*, quoted in B.L. Mijuskovic, *Loneliness in Philosophy, Psychology and Literature*, Assen, The Netherlands: Van Gorcum, 1979, p. 10.
17 *The Upanishads*, quoted in B.L. Mijuskovic, *op. cit.*, p. 11.
18 See H.B. Gibson, *The Emotional and Sexual Lives of Older People: a Manual for Professionals*, London: Chapman & Hall, Chapter 9, 1992.
19 Existential loneliness has been the subject of books by a number of well-known writers such as Albert Camus and Jean-Paul Sartre, whose ideas are best known to the reading public via their novels. Their philosophical works are rather hard going for the average reader. These writers see little value in the endeavours of the various brands of psychotherapy to overcome existential loneliness.
20 E.M. Forster, *A Passage to India*, Harmondsworth: Penguin Books, 1936, p. 195.
21 *Ibid.*, p. 147.
22 St Augustine, *Confessions*, Book 1, Chapter 1, quoted in Ronald Rolheiser, *op. cit.*, p. 117.
23 Ronald Rolheiser, *op. cit.*, p. 65.
24 A.N. Wilson, *Against Religion: Why We Should Try to Live without It*, London: Chatto and Windus, 1991, p. 1.
25 Clark Moustakis, *Loneliness and Love*, New Jersey: Prentice Hall, 1972.

2 The Problems of Later Life

1 See the Introduction to Alex Comfort, *A Good Age*, Revised Edition, London: Pan Books, 1990.
2 This matter is discussed by Alan Walker, 'The Benefits of Old Age', in Evelyn McEwan (ed.), *Age: the Unrecognized Discrimination*, London: Age Concern England, 1990, pp. 62–3.
3 Most of the physical effects of ageing are dealt with very fully in the various chapters of J. Grimley Evans (ed.), *Health and Function in the Third Age*, London: Nuffield Provincial Hospitals Trust, 1992.
4 Louis Harris and Associates, *The Myth and Realities of Aging in America*, Washington, DC: National Council on Aging, 1975.

5 V.L. Bengston, *The Social Psychology of Aging*, Indianapolis, NY: Bobbs Merrill, p. 27.
6 *Alzheimer's Disease – What Is It?* Information Sheet 1, London: Alzheimer's Disease Society, 1995.
7 S.H. Zarit and A.B. Edwards, 'Family Caregiving: Research and Clinical Intervention', in R.T. Woods (ed.), *Handbook of the Clinical Psychology of Ageing*, Chichester: Wiley, 1997.
8 *Ibid.*, p. 363.
9 'In England Now', *The Lancet*, 18 January 1986, p. 147.
10 David H. Clark, 'Living in the Third Age: a Report on U3A Discussion Groups 1984–1988', in V. Futerman (ed.), *Into the 21st Century*, University of the Third Age in Cambridge, 1989, p. 36.
11 Alex Comfort, *op. cit.*, pp. 21–2.
12 Simone de Beauvoir, *Old Age*, tr. Patrick O'Brian, London: André Deutsch and Weidenfeld & Nicolson, 1992, p. 7.
13 R.T. Woods, 'Institutional Care', in R.T. Woods, *op. cit.*
14 See P. Naylor, 'In Praise of Older Workers', *Personnel Management*, November 1987, pp. 44–8.
15 For a brief history of the introduction of pensions in the UK, see F. Lazco and C. Phillipson, 'Defending the Right to Work', in Evelyn McEwan, *op. cit.*
16 Alex Comfort, *op. cit.*, pp. 131–2.
17 E. Cumming and W. Henry, *Growing Old: the Process of Disengagement*, New York: Basic Books, 1961.
18 N. Wells and C. Freer, (eds.), *The Ageing Population: Burden or Challenge*, London: Macmillan, 1988.
19 Simone de Beauvoir, *op. cit.*, p. 59.
20 William Osler, quoted by W. Graebner, *A History of Retirement*, New Haven: Yale University Press, 1980, pp. 4–5.
21 Donald Gould, 'Death by Decree', *New Scientist*, Vol. 114, 1987, p. 65.
22 There is extensive recent literature on this subject, much of it reviewed by Peter Laslett, *A Fresh Map of Life*, Second Edition, London: Macmillan, 1996.
23 Age Concern England, Retirement Form: *Responses to Questionnaire*, (Unpublished) London, 1989.
24 Eric Midwinter, 'Your Country Doesn't Need You', in Evelyn McEwan, *op. cit.*
25 Fionnuala McKiernan, 'Bereavement and Attitudes to Death', in R.T. Woods, *op. cit.*, 1996.

3 The Measurement of Loneliness

1 Daniel Russell, 'The Measurement of Loneliness', in L.A. Peplau and D. Perlman (eds.), *Loneliness: a Sourcebook of Current Theory, Research and Therapy*, Chichester: Wiley, 1982.
2 Daniel Perlman and P. Joshi, 'The Revelation of Loneliness', in

M. Hojat and R. Crandell (eds), *Loneliness: Theory, Research and Therapy*, London: Sage, 1989.

3   For details of the University of the Third Age, see Appendix A.
4   For details of the statistical technique, see *Psychometrika* 1953, Vol. 18, p. 118.
5   J.J. Lynch, *The Broken Heart: the Medical Consequences of Loneliness* New York: Basic Books, 1977.
6   Mildred Baxter, 'Self-Reported Health', in *The Health and Lifestyle Survey*, London: Health Promotion Research Trust, 1987.
7   J.M. Mossey and E. Shapiro, 'Self-Rated Health: a Predictor of Mortality among the Elderly', *American Journal of Public Health*, 1982, Vol. 72, pp. 800–8.
8   For a discussion of the problems of self-rated health, see George L. Maddox, 'Self-Assessment of Health States', in E. Palmore (ed.), *Normal Aging*, Durham NC: Duke University Press, 1972.
9   Mildred Baxter, *op. cit.*
10   Considerable detail about such relationships is given in H.B. Gibson, *Love in Later Life*, London: Peter Owen, 1997.
11   L.A. Peplau, M. Miceli and B. Morasch, 'Loneliness and Self-Evaluation', in L.A. Peplau and D. Perlman (eds), *op. cit.*

4   Loneliness in Literature

1   Robert Potter, *The English Morality Play*, London: Routledge and Kegan Paul, 1975.
2   G. Cooper and C. Worthan (eds), *The Summoning of Everyman*, Nederlands, W.A.: Western Australian Press, 1980.
3   Francis A. Wood, 'Elckerlijc – Everyman: the Question of Priority', *Modern Philology*, 1910, Vol. 8, 279–302.
4   St Thomas Aquinas, *Summa Theologia*, quoted in Ronald Rolheiser, *The Restless Heart*, London: Hodder and Stoughton, 1979.
5   Francesco Petrarca, *De Vita Solitaria*, Leiden: Universitaire Pers Leiden, 1990.
6   William Shakespeare, 'As You Like It', III.2, in *The Works of William Shakespeare*, London: Basil Blackwell, 1947.
7   *Ibid.*, III.3.
8   Hardin Craig, 'Morality Plays and Elizabethan Drama', *Shakespeare Quarterly*, 1950, Vol. 1, pp. 64–72.
9   William Shakespeare, 'King Lear', V. 3, *op. cit.*
10   Ernest Jones, *The Problem of Hamlet and the Oedipus Complex*, London: Vision Press, 1947.
11   Jean-Paul Sartre, *Les Mains Sales*, London: Routledge, 1988.
12   M.E. Novak, 'Swift and Defoe, or how Contempt Breeds Familiarity and a Degree of Influence', in H.J. Reed and H.I. Vienken (eds), *Proceedings of the 1st Munster Symposium of Jonathan Swift*, Munchen: Verlag, 1985.

13 Martin Green, *The Robinson Crusoe Story*, University Park, Pennsylvania: Pennsylvania State Press, 1990.
14 Quoted in Paula R. Backscheider, *Daniel Defoe: His Life*, Baltimore: Johns Hopkins University Press, 1989, p. 33.
15 Aristotle, *Politics*, quoted in B.L. Mijuskovic, *Loneliness in Philosophy, Psychology and Literature*, Assen, The Netherlands; Van Gorcum, 1979, p. 3.
16 Jonathan Swift, *Gulliver's Travels*, Harmondsworth: Penguin Books, 1967, p. 258.
17 Alfred Tennyson, 'Tithonus', in *The Poems of Tennyson*, Harmondsworth: Penguin Books, 1991.
18 L. McPike, *Dostoevsky and Dickens*, London: George Prior and Sons, 1981.
19 Franz Kafka, *The Trial*, tr. W. and E. Muir, London: Secker and Warburg, 1956.
20 Charles Dickens, 'Autobiographical Note' quoted in Edgar Johnson, *Charles Dickens: His Tragedy and Triumph*, New York: Simon Schuster, 1952, p. 44.
21 Natalie McKnight, *Idiots, Madmen and Other Prisoners in Dickens*, New York: St Martin's Press, 1993, p. 23.
22 Charles Dickens, 'Tom Tiddler's Ground', in *Christmas Stories*, London: Chapman and Hall, (no date) p. 300.
23 *Idem*, *Dombey and Son*, London: Chapman and Hall (n.d.), p. 98.
24 *Idem*, *Oliver Twist*, London: Chapman and Hall (n.d.), p. 87.
25 *Idem*, *Nicholas Nickleby*, London: Chapman and Hall (n.d.), p. 780.
26 Bill Bytheway, *Ageism*, Buckingham: Open University Press, 1995.
27 Adam Gillon, *Joseph Conrad*, Boston: Twain Publishers, 1982.
28 B.L. Mijuskovic, *op. cit.*, p. 36.
29 Joseph Conrad, *Loneliness in Philosophy, Psychology and Literature*, Assen, Netherlands: Van Gorcum, 1979. 'Author's Note', in *Nostromo*, London: J.M. Dent, 1962.
30 Evelyn Waugh, *A Handful of Dust*, London: Chapman and Hall, 1964.
31 Roland Reinart, *Evelyn Waugh: 'A Handful of Dust'*, Harlow: Longman York Press, 1984, p. 104.
32 Roger Tennant, *Joseph Conrad: a Biography*, London: Sheldon Press, 1981.
33 Bertrand Russell, *The Autobiography of Bertrand Russell*, Vol. 1, London: Allen and Unwin, 1967, pp. 207–8.
34 Joseph Conrad, 'Amy Foster', in Cedric Watts (ed.), *Typhoon and Other Tales*, Oxford: Oxford University Press, 1986.
35 *Idem*, *Heart of Darkness*, London: Hodder and Stoughton, 1990.
36 *Idem*, *Lord Jim*, London: J.M. Dent, 1946.
37 *Idem*, 'Typhoon', in Cedric Watts (ed.), *op. cit.*
38 Graham Greene, *In Search of a Character: Two African Journals*, London: Bodley Head, 1961.

39 *Idem, The Heart of the Matter*, London: Heinemann, 1948.
40 *Idem, A Burnt-Out Case*, London: Heinemann, 1961.
41 Joseph Conrad, *Victory: an Island Tale*, Harmondsworth: Penguin Books, 1994.
42 Adam Gillon, *Joseph Conrad*, Boston: Twain Publications, 1992, p. 141.
43 Jeffrey Myers, *Homosexuality in Literature*, Montreal: McGill-Queens University, 1977.
44 Robert Hedges, 'Deep Fellowship: Homosexuality and Male Bonding in Life and Fiction of Joseph Conrad', *Journal of Homosexuality*, 1979, Vol. 4, pp. 379–93.
45 Radclyffe Hall, *The Well of Loneliness*, London: Hutchinson, 1986.
46 Tom Wakefield, *Mates*, London: Hutchinson, 1996.
47 George Orwell, *Animal Farm: a Fairy Story*, London: Secker and Warburg, 1945.
48 *Idem, Nineteen Eighty-Four*, London: Secker and Warburg, 1949.
49 Frederic Warburg, 'Review of *Nineteen Eighty-Four*', quoted in Bernard Crick, *George Orwell: a Life*, London: Secker and Warburg, 1980, p. 396.
50 George Orwell, quoted in Bernard Crick, *op. cit.*, p. 398.
51 *Idem*, 'Politics vs Literature: an Examination of *Gulliver's Travels*', *Polemic*, 1946, No. 5, September–October.
52 Patrick Reilly, *George Orwell: the Age's Adversary*, London: Macmillan, 1986.
53 George Orwell, 'Such, Such were the Joys', *Partisan Review*, 1952, September–October.
54 Peter Davison, *George Orwell: a Literary Life*, London: Macmillan, 1996.
55 The two post-Orwellian utopias are Aldous Huxley, *Island*, London: Flamingo, 1964, and B.F. Skinner, *Walden Two*, New York: Macmillan,1962.
56 For a discussion of later literature dealing with the lives of elderly people, see H.B. Gibson, *Love in Later Life*, Chapter 2, London: Peter Owen.

5 The Benefits of Solitude

1 See John Bartlett's *Familiar Quotations*, 15th Edition, London: Macmillan, 1980.
2 Jonathan Swift, 'The Faculties of Mind', in C. Van Doren (ed.), *The Portable Swift*, Harmondsworth: Penguin Books, 1977.
3 Laurence Sterne, *Correspondence*, No. 82, Dublin: R. Stewart, 1779.
4 Isaac D'Israeli, *Literary Character of Men of Genius*, Chapter 10, New York: J. Earburn, 1818.
5 James Thomson, 'Hymn to Solitude', in *The Poetical Works of James Thomson*, Philadelphia: Benjamin Johnson, 1804.

6 William Wordsworth, *A Poet's Epitaph*, London: Ginn & Co., 1932.
7 L.E. Hinkle and H.G. Wolff, 'Communist Interrogation and Indoctrination of "Enemies of the State"', *Archives of Neurology and Psychiatry*, 1956, Vol. 76, pp. 115–74.
8 Arthur Koestler, *Darkness at Noon*, London: Cape, 1940.
9 *Idem*, *Kaleidoscope*, London: Hutchinson, 1981, p. 208.
10 Edith Bone, *Seven Years Solitary*, London: Macmillan, 1957.
11 Christopher Burney, *Solitary Confinement*, London: Macmillan, 1984.
12 J.A.H. Murray, *A New English Dictionary*, Oxford: Clarendon Press, 1888.
13 W.D. Whitney, *The Century Dictionary*, New York: The Times Book Club, 1889.
14. H.J. Eysenck, *Eysenck on Extraversion*, London: Granada, 1973, p. 3.
15 H.J. Eysenck and S.B.G. Eysenck, *The Eysenck Personality Inventory*, London: University of London Press, 1965.
16 G.C. Drew, W.P. Colquhoun and H.A. Long, 'Effect of Small Doses of Alcohol on a Skill Resembling Driving', *M.R.C. Memorandum No. 38*, London: HMSO, 1959.
17 H.J. Eysenck, 'Personality and Ageing: an Exploratory Analysis', *Journal of Social Behavior and Personality*, 1987, Vol. 3, pp. 11–21.
18 S. Stuart-Hamilton, *The Psychology of Ageing*, Second Edition, Chapter 5, London: Jessica Kingsley, 1994.
19 R.L. Rubenstein, *Singular Paths: Old Men Living Alone*, New York: Columbia University Press, 1986.
20 L.A. Peplau, T.K. Bikson, K.S. Rook *et al.*, 'Being Old and Living Alone', in L.A. Peplau and D. Perlman (eds), *Loneliness*, New York: J. Wiley & Sons, 1982, p. 331.
21 *Ibid.*, p. 342.
22 There has been a long debate about the utility or inutility of psychotherapy as an effective therapeutic agent, which began with H.J. Eysenck, 'The Effects of Psychotherapy: an Evaluation', *Journal of Consulting Psychology*, 1952, Vol. 16, pp. 319–24. For a recent review, see Bob Potter, 'Psychotherapy and Religion', *The Raven*, 1997, Vol. 9, No. 3, pp. 266–82.
23 M. Miura and I. Usa, 'A Psychotherapy of Neurosis: Morita Therapy', *Yonago Acta Medica*, 1970, Vol. 14, pp. 1–17.
24 H.B. Gibson, 'Morita Therapy and Behaviour Therapy', *Behavior Research and Therapy*, 1974, Vol. 12, pp. 342–53.
25 Anthony Storr, *Solitude*, London: Fontana, 1989, pp. 29–30.
26 B. Ineichen, *Senile Dementia: Policy and Services*, London: Chapman & Hall, 1989.
27 Richard E. Byrd, *Alone*, London: Queen Anne Press, 1987, p. 122.
28 *Ibid.*, pp. 134–43.
29 Anon., *Unpublished Diary*, 1986.
30 Woodes Rogers, quoted by Pat Rogers, *Robinson Crusoe*, London: Allen & Unwin, 1979.

31 Richard Steele, quoted by Pat Rogers, *op. cit.*, p. 161.
32 Richard Byrd, quoted by Anthony Storr, *op. cit.*, p. 37.

6 Overcoming Loneliness

1 Alex Comfort, *A Good Age* (Revised Edition), London: Pan Books, 1990, p. 33.
2 Wendy and Sally Greengross, *Living, Loving and Ageing*, Mitcham: Age Concern England, 1989, p. 93.
3 M.E. Swiger, M.B. Bowers and S. Fleck, 'Grieving and Unplanned Pregnancy', *Psychiatry*, Vol. 39, pp. 72–9.
4 G. Gorer, *Death, Grief and Mourning in Contemporary Britain*, London: Cresset Press, 1965.
5 C. Murray Parkes, *Bereavement: Studies in Grief in Adult Life*, Harmondsworth: Penguin Books, 1988, pp. 178–9.
6 *Idem*, pp. 50–107.
7 P. Speck, *Loss and Grief in Medicine*, London: Baillière Tindall, 1978.
8 B. Cook and S.G. Phillips, *Loss and Bereavement*, London: Austin Cornish Books, 1988.
9 Margaret Savory, 'Depression', in J. Grimley Evans (ed.), *Health and Function in the Third Age*, London: Nuffield Provincial Hospital Trust, 1993.
10 H.B. Gibson, *Love in Later Life*, London: Peter Owen, 1997.
11 Quoted from H.B. Gibson, 'Emotional and Sexual Re-Adjustment in Later Life', in S. Arber and M. Evandrou (eds), *Ageing, Independence and the Life Course*, London: Jessica Kingsley, 1993, pp. 117–18.
12 R.E. Butler and M.I. Lewis, *Love and Sex After 60*, Revised edition, New York: Harper & Row, 1988, pp. 113–14.
13 See Chapter 5 in H.B. Gibson, *The Emotional and Sexual Lives of Older People: a Manual for Professionals*, Chapman and Hall, 1992.
14 Arthur Miller, *Death of a Salesman*, London: Heinemann, 1968.
15 C.R. Rogers, 'The Lonely Person and His Experiences in an Encounter Group', in *Carl Rogers on Encounter Groups*, New York: Harper & Row, 1973.
16 *Ibid.*, p. 119.
17 H.B. Gibson, *On the Tip of Your Tongue: Your Memory in Later Life*, London: Third Age Press, 1996.
18 *Ibid.*, pp. 126–7.
19 Quoted from A.E. Kahn (ed.), *Joys and Sorrows: Reflections by Pablo Casals*, New York: Simon Schuster, 1970, pp. 15–17.
20 L.A. Peplau *et al.*, 'Being Old and Living Alone', in L.A. Peplau and D. Perlman (eds), *A Sourcebook of Current Theory, Research and Therapy*, Chichester: J. Wiley & Sons, 1982, p. 334.
21 C. Murray Parkes, *Bereavement*, Harmondsworth: Penguin Books, 1986, p. 178.
22 B.D. Starr and M.B. Weiner, *The Starr–Weiner Report on Sex and Sexu-*

ality in the Mature Years, New York: McGraw Hill, 1971.

23 A. Assister and A. Carol, *Bad Girls and Dirty Pictures: the Challenge to Reclaim Feminism*, London: Pluto Press, 1993, p. 1.

24 *Ibid.*, p. 5.

25 The Hen Co-op, *Growing Old Disgracefully*, London: Piatkus, 1993.

26 Geoffrey Chaucer, 'The Prologe of the Wyves Tale of Bathe', in W.W. Skeate (ed.), *The Canterbury Tales*, London: Oxford University Press, 1940.

Appendix A: Useful Addresses for Older People

Age Action Ireland
114–116 Pearce Street
Dublin 2
Ireland   3531 677 9892
A network of providers promoting better policies and services for older people.

Age Concern England
1258 London Road
London SW16 4EJ   0181 679 8000

Age Concern Northern Ireland
3 Lower Crescent
Belfast BT7 1NR   01232 245 729

Age Concern Scotland
113 Rose Street
Edinburgh EH2 3DT   0131 220 3345

Age Concern Wales,
1 Cathedral Road
Cardiff CF1 9SD   01222 371566
Age Concern offers a wide range of services to older people and has numerous groups throughout the UK.

Alzheimer's Disease Society
10 Greencoat Place
London SWIP 1PH   0171 306 0606
Provides information and support for carers during illness and after death of the patient.

British Association for the Hard of Hearing
7–11 Armstrong Road
London W3 7JL   0181 743 1110
Organizes clubs throughout the country for people with all degrees of hearing loss. Some clubs run lip-reading classes.

Carers' National Association
20 Glasshouse Yard
London EC1   0171 490 8818
They provide information, advice and support for carers, including those
who have been recently bereaved.

Civil Service Fellowship
1B Deal's Gateway
Blackheath Road
London SE10 8BW   0181 691 7411
Offers a range of services for widows and widowers associated with the
Civil Service.

CRUSE Bereavement Care
Cruse House
126 Sheen Road
Richmond
Surrey TW9 1UR   0181 940 4818/0181 332 7227 (Helpline)
CRUSE offers help to all bereaved people through counselling, informa-
tion and social support groups, and has about 200 branches nationwide.

The Disabled Living Centres
Winchester House
11 Cranmer Road
London SW9 6EJ   0171 820 0567
They aim to enable disabled and elderly people to get accurate and impartial
information and advice about products designed to help daily living.
They can put you in touch with your nearest centre.

Forces Help Society
122 Brompton Road
London SW3 IJE   0171 589 3243
Offers a range of services for widows and widowers associated with the
Forces.

Gay Bereavement Project
Vaughan M. Williams Centre
Colindale Hospital
London NW9 5GH   0181 455 8894
Telephone counselling for those bereaved by a same-sex partner.

Growing Old Disgracefully Network
c/o Piatkus Books
5 Windmill Street
London W1P 1HF

This is a national network through which older women can be put in touch with other women in their area who wish to meet socially as described in the book *Growing Old Disgracefully*. (See Appendix B.) Apply by post and enclose a s.a.e.

Help the Aged
St James Walk
Clerkenwell Green
London ECIR OBE   0171 253 0253
They have a Freephone advice and information line (0800 650 065) for senior citizens and provide a range of free advice leaflets on health issues, housing, home safety and money matters. They also run residential homes and provide sheltered housing.

Jewish Bereavement Counselling Service
1 Cyprus Gardens
London N3 1SP   0181 349 0839
                       0171 387 4300 (24-hour answering phone).
They will send trained volunteer counsellors to the bereaved; they operate in Greater London but can refer people to other projects and individuals elsewhere.

National Association Bereavement Services
20 Norton Folgate
London El 6DB
Publishes a Directory of Bereavement and Loss Services.

National Association of Widows/Widowers Advisory Trust
54–57 Allison Street
Digbeth
Birmingham B5 5TH   0121 6543 8348
Advice, information and support is offered to all widows and widowers, their families and friends. There are local branches, contact lists and a pen club.

National Council for the Divorced and Separated
PO Box 519
Leicester LEZ 3ZE   0116 2700 595
They are concerned with the interests and welfare of divorced and separated people, with 100 branches and 10 000 members nationwide. Inquirers should include a s.a.e.

National Osteoporosis Society
PO Box 10
Radstock
Bath BA3 3YB

The Society provides advice and information through a national helpline and network of local groups, and raises funds for research. Please send a s.a.e.

Northern Ireland Widows' Association
137 University Street
Belfast BT7 1HP   01232 228 263
This has a local network offering information, advice and support.

Parkinson's Disease Society
22 Upper Woburn Place
London WC1   0171 383 3513
This society offers information, publications and support concerning all aspects of Parkinson's Disease. They produce a *Newsletter*, fund research, and have over 220 local branches.

The Royal National Institute for the Blind
224 Great Portland Street
London W1N 6AA   0171 388 1266
The RNIB provides free information packs and a wide range of services for people suffering from all degrees of visual impairment. Details of your local branch will be supplied on request.

The Royal National Institute for Deaf People
19–23 Featherstone Street
London EC1Y 8SL   0171 296 8000 (voice)
                        0171 296 8001 (text)
The RNID is the largest voluntary organization in the UK representing people with hearing disabilities. They provide information, residential care, training, specialist telephone services, communication support and assistive devices.

The Samaritans
10 The Grove
Slough SLI 1QP   01753 532 713
They offer a 24-hour emotional befriending service and have numerous local branches. See telephone directory for your local centre.

The Stroke Association
CHSA House
123–127 Whitecross Street
London EC1Y 8JJ   0171 490 7999
They give help and advice to people who have suffered a stroke and to their families. They offer a community service.

The University of the Third Age
26 Harrison Street
London WC1H 8JG   0171 837 8838
There are over 300 local branches and the membership is in excess of
50 000. The primary aim is to provide low-cost educational and social
opportunities to people in later life who are no longer in full employment. There are no entrance qualifications and the branches are
autonomous, depending on the self-help principle, without the aid of
paid tutors.

The Voluntary Euthanasia Society
13 Prince of Wales Terrace
London W8 5PG   0171 937 7770
The Society's principal object is to make it legal for an adult person
suffering severe distress from an incurable illness to receive medical help
to die at their own considered request. They also issue an advance di-
rective allowing you to reject unwanted and futile treatment which merely
prolongs the process of dying.

War Widows' Association of Great Britain
81 Gargrave Road
Skipton BD23 1QN   01756 793 719
They are chiefly concerned with those who are in financial difficulties.

Appendix B: Useful Books

Gerry Bennett, *Alzheimer's Disease and Other Dementias*, London: Vermillion, 1994.

Pat Blair, *Know Your Medicines*, London: Age Concern, 1991.

British Medical Association, *Complementary Medicine*, Oxford: Oxford University Press, 1994.

Jane Brotchie, *Caring for Someone Who Has Dementia*, London: Age Concern, 1995.

R.E. Butler and M.I. Lewis, *Love and Sex after 60*, Revised Edition, New York: Harper and Row, 1988.

Alex Comfort, *A Good Age*, London: Pan Books, 1989.

Philip Coyne and Penny Mares, *Caring for Someone Who Has a Stroke*, London: Age Concern, 1995.

H.B. Gibson, *On the Tip of Your Tongue: Your Memory in Later Life*, London: Third Age Press, 1995.

H.B. Gibson, *A Little of What You Fancy Does You Good: Your Health in Later Life*, London: Third Age Press, 1997.

Tony Gibson, *Love, Sex and Power in Later Life*, London: Freedom Press, 1992.

Margaret Graham, *Keep Moving, Keep Young: Gentle Yoga Exercises for the Elderly*, London: Unwin Hyman, 1988.

Wendy and Sally Greengross, *Living, Loving and Ageing*, London: Age Concern, 1992.

Hamilton Hall, *Be Your Own Back Doctor*, London: Grafton Books, 1987.

The Hen Co-op, *Growing Old Disgracefully*, London: Piatkus, 1993.

In Touch Publishing, *In Touch Handbook: a Comprehensive Guide to All Services, Benefits and Equipment Available to Blind and Partially Sighted People*, London: In Touch Publishing, 1995.

In Touch Publishing, *101 Practical Hints For Living with Poor Sight*, London: In Touch Publishing, 1995.

Lorraine Jeffrey, *Hearing Loss and Tinnitus*, London: Ward Lock, 1995.

Leslie Kenton, *The New Ageless Ageing*, London: Arrow Books, 1995.

J.J. Lynch, *The Broken Heart: the Medical Consequences of Loneliness*, New York: Basic Books, 1977.

J.A. Muir Gray and Pat Blair, *Your Health in Retirement*, London: Age Concern, 1990.

C. Murray Parkes, *Bereavement: Studies in Grief in Adult Life*, Harmondsworth: Penguin Books, 1988.

Barbara Robertson, *Fitness for the Over 50s*, Wellingborough: Patrick Stephens, 1988.

Royal National Institute for the Blind, *You and Your Sight: Living With a Sight Problem*, London: RNIB, 1994.

George Target, *Your Arthritic Hip and You*, London: Sheldon Press, 1987.

Keith Thompson, *Caring for an Elderly Relative*, London: Martin Dunitz, 1986.

J. Tunstall, *Old and Alone*, London: Routledge and Kegan Paul, 1967.

Richard Villar, *Hip Replacement: a Patient's Guide to Surgery and Recovery*, London: Thorson's Health Series, 1995.

Geoffrey Webb and June Copeman, *The Nutrition of Older Adults*, London: Edward Arnold, 1996.

Fiona van Zwanenberg and Jackie Corder, *Taking Care of Yourself*, Bicester: Winslow Press, 1990.

Index

affluence/poverty, 12
Age Concern, xvi, 28, 37
ageism, 34–7
Alzheimer's disease, 24
Aristophanes, 14–15
Aristotle, 65, 72
Assister, A., 127–8

Bengston, Vern, 23
bereavement, 24, 105–6,
   109–17
Bergson, Henri, xv
Beta coefficient, 41, 43
biogenic problems, 21–5
Blake, William, 106
blindness, 11–12, 23
Bone, Edith, 93–4, 104
brainwashing, 93
Brontë, Emily, 62
Burney, Christopher, 94
Butler, Robert, 119–20
Butler, Samuel, 74
Byrd, Richard, 104–5, 108

Camus, Albert, 62
carers, 24–5
Carol, A., 127–8
Cervantes, Miguel, 62
Ching Chow, xv, 96
Chaucer, Geoffrey, 129
church attendance, 127
Coleridge, S.T., 62
Comfort, Alex, 11–12, 27, 111
Communist Party, 89, 94
Conrad, Joseph, xv, 78–85
Cook, B., 115
creativity, 13, 19

Davison, Peter, 89
deafness, 11–12, 23–4

De Beauvoir, Simone, 28, 33
Defoe, Daniel, xv, 69–72, 106
dementia, 24
depression, 5, 104–5, 108, 113,
   116–17
De Quincy, Thomas, 62
deterioration, natural, 22, 62
Dickens, Charles, xv, 8, 61,
   74–8, 88
disengagement, 29–32, 34–7
D'Israel, Isaac, 91
divorce/separation, 45–6, 109,
   120–1
Dostoevsky, Fedor, 62, 75

Ecclesiastes, 19–20
Eliot, George, 62
Eliot, T.S., 62
enshrinement, 119–20
Eskimo tradition, 33
existential loneliness, 16–17,
   19–20
extended families, 7
Eysenck, Hans, 96–8
Eysenck Personality Inventory,
   97–8

fantasy loneliness, 8–10
feminism, 89, 127–8
Finch, Edith, 3
Forster, E.M., 16–17, 101
Forster, John, 75

Gillon, Adam, 79, 84
Golding, William, xv, 62
Gotesky, Reuben, 4–6
Gould, Donald, 35
'granny-dumping', 33–4
Greek culture, 14
Greene, Graham, 78–83, 85

Greengross, Wendy and Sally, 112–13
Hall, Radcliffe, 85
handicapped people, 10–11
Hardy, Thomas, xv, 62, 82
hearing-aids, 12
Hemingway, Ernest, 62
Hindu tradition, 15, 32–3
homosexuality, 85
Horney, Karen, 102
Huxley, Aldous, 90

Ibsen, Henrik, 65, 83
Ineichin, Bernard, 103
institutions, 28–9
interviewing, 39–40
introversion-extraversion, 95–8

Japanese tradition, 33, 101
Jewish tradition, 14
Jones, Ernest, 69
Joyce, James, xv, 62
Jung, Carl, 95–7

Kafka, Franz, 62
Kant, Immanuel, xv
Kierkegaard, Søren, xv
Koestler, Arthur, 93

Laslett, Peter, xvi
Lear, Edward, 62
Lewis, M.I., 119–20
Lodge, David, xv
London, Jack, 62
Loneliness
  measure of differences in: age, 43–4; gender, 44; marital status, 45–6, 50; health status, 53–4
  reasons for, 55
  types: personal characteristic, 2–3; physical, 4; restless, 18; solitude, 6; state/trait, xiii, 2, 4
Lodge, David, xv
Louis Harris poll, 20

love, importance post-bereavement, 117–19

Malory, Sir Thomas, 65
Mann, Thomas, 62
Maugham, W.S., xv, 62
medieval period, 62
Melville, Herman, 62, 79
'mental defectives', 10–11
Midwinter, Eric, 37
Mijuskovic, Ben, 79
mobility problems, 12, 23
morality plays, 63
Morita, Shoma, 101–2
Morita therapy, 101–2
mourning, 115
Moustakis, Clark, 19
Murray, J.A.H., 106
Myers, Jeffrey, 85
mystery plays, 62–3

'new old', 127

'Oak Mountain' song, 33
Orwell, George, 74, 86–90
Osler, William, 34–5

Pankhurst, Emmeline, 128
Parkes, Murray, 113–15, 127
pensions, 21, 30
Peplau, L.A., 99–100, 126–7
Pepys, Samuel, xiii
Petrarch (Petraca, Francesco), 65
Phillips, S.G., 115
physical fitness, 23
Plato, 14
Proust, Marcel, xv, 62
public transport, 12

questionnaires, 39–40, 42

racism, 35–6
religion, 16–18
Reinert, Roland, 80
'renaissance man', 66
retirement, 9, 29–32, 37, 121–5

Rogers, Woodes, 106–7
Rolheiser, Ronald, 8, 18
Rubenstein, R.L., 98
Russell, Bertrand, 2–4, 18, 81–2

St Thomas Aquinas, 72
Salinger, J.D., 61
Sartre, Jean-Paul, 69
Schopenhauer, Arthur, xv
Selkirk, Alexander, 106–8
sexist discrimination, 36
sexuality, 14, 25, 85, 112–13
Shakespeare, William, 66–9
shivah, 113–14
Skinner, B.F., 90
sociogenic problems, 21, 25–32
soap operas, 7
solitary living, 98–101
solitude
    a form of loneliness, 13
    imposed, 92–5
    in later life, 95
    personal growth, 104
    temperamental, 95–8
status, loss of, 121–2
spinsterhood, 49
sport, 6–7
Steele, Richard, 106–7
stereotype of elderly, 27
Sterne, Laurence, 91
Storr, Anthony, 103
Struldbruggs, 5, 73, 88
Stuart-Hamilton, S., 98

Summoning of Everyman, 63–4
Swift, Jonathan, xv, 5, 69, 71–3, 91
Swigar, M.E., 113

Tennyson, Alfred, 73–4
Third Age, xvi, 20, 22, 23, 26
Thompson, James, 91–2
Tithonus, 73–4
Tolstoy, Leo, 75
Trollope, Anthony, 34
Twain, Mark, 51

Upanishads, 15
Usa Gen-yu, 102
University of the Third Age, xvi,
    28, 40–1, 125–6

'village idiot', 10

Wakefield, Tom, 85
Warburg, Frederic, 86
Waugh, Evelyn, 80
Weiner, M.B., 127
Wendell Holmes, Oliver, 13
Wilde, Oscar, 74
Wilson, A.N., 18
Whitney, W.D., 96
Wolfe, Thomas, xv, 62
women's suffrage movement, 128
workhouses, 29

Yeats, W.B., 2, 62